Neuro-Fuzzy Equalizers for Mobile Cellular Channels

Neuro-Fuzzy Equalizers for Mobile Cellular Channels

K. C. Raveendranathan, Ph. D.

CRC Press
Taylor & Francis Group
Boca Raton   London   New York

CRC Press is an imprint of the
Taylor & Francis Group, an **informa** business

MATLAB® is a trademark of The MathWorks, Inc. and is used with permission. The MathWorks does not warrant the accuracy of the text or exercises in this book. This book's use or discussion of MATLAB® software or related products does not constitute endorsement or sponsorship by The MathWorks of a particular pedagogical approach or particular use of the MATLAB® software.

CRC Press
Taylor & Francis Group
6000 Broken Sound Parkway NW, Suite 300
Boca Raton, FL 33487-2742

First issued in paperback 2017

© 2014 by Taylor & Francis Group, LLC
CRC Press is an imprint of Taylor & Francis Group, an Informa business

No claim to original U.S. Government works

Version Date: 20130703

ISBN 13: 978-1-138-07660-0 (pbk)
ISBN 13: 978-1-4665-8152-4 (hbk)

Visit the Taylor & Francis Web site at
http://www.taylorandfrancis.com

and the CRC Press Web site at
http://www.crcpress.com

To

my wife Sobha
and daughters Anuja & Aabha

Contents

List of Figures

List of Tables

Preface

The purpose of a communication system is to transfer information between two separate points over some medium in the presence of disturbances or distortions such as noise and dispersion. This distortion is manifested in the time domain as pulse dispersion and is labeled as Inter-Symbol Interference (ISI). As data rates increase in modern digital communication systems, ISI becomes an inevitable consequence of the dispersive nature of band-limited propagation channels. The receiver must include an equalizer to mitigate the effects of ISI. Thus, an equalizer undoes the distortion that the signal is subjected to while it propagates through the channel. Needless to say, equalizers are present in all forms of communication systems: from Plain Old Telephones Systems (POTS) to Co-axial communication systems, to RF and Microwave communication systems, to Optical Fiber communication systems, and to wireless mobile communication systems. The function of the equalizer is to combat the ISI and to utilize the available spectrum most efficiently. Equalizers are cascaded to almost all kinds of channels, right from telephone lines to radio and optical fiber channels, to make the channel performance optimal. Ideally, an equalizer, when cascaded to the end of a channel, will make it behave like an ideal channel, one which will not distort the signals in any manner. In the case of mobile cellular channels, which are generally considered to be Linear Time Variant (LTV), the design of equalizers is *not a trivial problem*. Moreover, the above said channel has certain uncertainties in its behavior, which need to be tackled in the equalizer design. The Co-Channel Interference (CCI) due to frequency reuse and Adjacent Channel Interference (ACI) due to spectral leakage both contribute to the reduction in the overall Signal-to-Interference-Noise-Ratio (SINR) in mobile cellular channels.

In applications in which the Channel Impulse Response (CIR) is unknown and no training sequence is available, the equalizer must be computed/updated blindly from the received signal and knowledge of the statistics of the data source alone. A common approach in continuous transmission systems is to blindly update a Linear Equalizer (LE) using the Constant Modulus Algorithm (CMA), and then switch to a Decision Directed (DD) mode when the Symbol Error Rate (SER) is low enough. Switching to a DD-based decision feedback equalizer (DFE) is also possible and desirable.

Modeling and simulation of mobile cellular channels have been successfully carried out by several researchers. Various interference patterns including Ricean/Rayleigh fading, co-channel and adjacent-channel interferences can be found in the literature. This book is intended to discuss the modeling of the mobile cellular channel used in an indoor environment, where the channel can be taken to be of the slow fading type. The study is focused to consider the noise contributions from various sources, when they fall within the spectrum of the frequencies used in cellular telephony, and then to design an equalizer which will mitigate the noise present due to CCI and ACI. When the channel over which data is sent is unknown, which is common, one must employ adaptive equalization. The DFE is one such adaptive equalizer. It is known that the DFE generally outperforms the LE for the same hardware complexity. Further, as indicated earlier, when the channel characteristics show Rayleigh/Ricean fading (due to the presence of a multipath), ACI and CCI, realization of equalizers based on neuro-fuzzy techniques seems to be the most appropriate option for the mobile cellular channel.

Linear space-time equalization is shown to be effective in coping with the complicated propagation conditions for wireless broadband communication in an industrial indoor environment. This is demonstrated by realistic simulations that use a real channel sounder for modeling the influence of the radio channel. Industrial indoor environments like large factory halls typically show a complicated radio channel because of the presence of many reflecting objects. This results in wide delay spreads and a considerably changing channel for a moving mobile unit. There exist a number of options to overcome the difficulties of heavy multipath propagation.

In this book, the mobile channel is modeled as a linear time variant channel. Further, the issues in the design of the neuro-fuzzy channel equalizer to null the effects of fading are investigated. One of the objectives of this work is to establish the fact that, within an acceptable bound, the mobile cellular channel is LTV. *Another major objective of the book is to investigate the suitability of neuro-fuzzy models as applicable to the analysis and design of mobile cellular channel equalizers.* Three solutions to the channel equalizer problem are investigated in this work. First, a type-2 Fuzzy Adaptive Filter (FAF) for the above purpose is considered. Simulations show that it performs better than a type-1 FAF or Neural Network Classifier (NNC) equalizer. Then the use of an Adaptive Network-Based Fuzzy Inference System (ANFIS) is investigated. Last, a Compensatory Neuro-Fuzzy Filter (CNFF) for channel equalization is considered. Subsequently, an attempt is made to bring the various equalizer realizations in the study under the generic framework of a radial basis function (RBF) neural network. Further, a novel modular approach for the simulation and design of equalizers for Nonlinear Time-Variant (NLTV) channels is proposed. A suitable model for an Ultra-Wide Band (UWB) channel and its equalization is the last goal.

The contributions of this work are the establishment of the fact that the mobile cellular channel can indeed be modeled as an *LTV channel*, in general, with a *Rayleigh distribution* for the channel coefficients. It is shown that FAF-, CNFF-, and ANFIS-based equalizers are capable of achieving desired SNR in the presence of CCI and ACI. It is also shown that the channel equalizers based on type-2 FAF, CNFF, and ANFIS could be brought under the generic framework of RBF Neural Networks. A detailed performance evaluation of the equalizers is made. And, finally, a modular approach for the simulation and modeling of NLTV channels is proposed. In the beginning it was mentioned that mobile channels are considered to be LTV. However, when the transmitter stages are driven to their nonlinear regions, the channel needs to be modeled as nonlinear (to account for the nonlinearities thus introduced to the transmitted signal) and Time Variant (NLTV). The modular approach in combating ISI is to cascade an adaptive nonlinear preprocessor filter and linear adaptive equalizer, which simplifies the equalizer design. It is also shown that the ANFIS model can be successfully adapted to equalization of UWB channels.

The book is organized thus. Chapter 1 gives a brief introduction to channel equalizers. Chapter 2 begins with a study of the nature of mobile cellular channels with regard to the frequency reuse and the resulting CCI. Several channel models available for mobile cellular channels are considered and the one best suited for our system is selected. It is established that the mobile indoor channel is a Rayleigh fading channel. The channel equalization problem is presented. It is succeeded by a study of various equalizers for mobile cellular channels. It starts with a discussion on conventional equalizers like LE and DFE using a simple LMS algorithm and transversal equalizers. Then channel equalization with neural networks and fuzzy logic is discussed, and various equalizers are classified.

In Chapter 3, the concept of fuzzy logic controllers in noise cancellation problems is considered in detail. This being a relatively new branch of study, the fundamental concepts of neuro-fuzzy systems are given and FAFs that are used in a variety of applications are discussed. Type-2 fuzzy sets are introduced and their advantages in overcoming certain short falls of conventional fuzzy sets (type-1 fuzzy sets) in dealing with real life problems are discussed. The performance of the type-2 fuzzy adaptive filter (FAF-II) is compared with the type-1 fuzzy adaptive filter (FAF-I) and NNC for the same purpose.

In Chapter 4, the ANFIS-based channel equalizer for mobile cellular and UWB channels is treated in detail. Contemporary literature provides sufficient information on the statistical properties of mobile cellular channels. The training of the ANFIS-based channel equalizer is based on the above mentioned statistics. As the ANFIS synergically combines the learning capability of the neural network and the decision-making capability (in presence of noise) of the fuzzy system, it can outperform the FAF as well as neural network-based channel equalizers.

In Chapter 5, the CNFF is considered for equalization. It is shown that the performance of the CNFF is well suited for the equalization of nonlinear channels.

A generic framework of RBF neural networks for the three filter structures developed in previous chapters (Chapters 3, 4, and 5) is established in Chapter 6. It is shown that under certain conditions, the type-2 FAF (FAF-II), CNFF, and ANFIS behave as an RBF network. A novel modular approach for the simulation and design of nonlinear time-variant channels is proposed in Chapter 7. It is established that the modular approach is the most appropriate one to model transmitter nonlinearities.

In Chapter 8, we discuss orthogonal frequency division multiplexing (OFDM) and spatial diversity techniques. Equalizers used for OFDM channels are considered here. In Chapter 9, the work done is summarized and venues for further research are explored. As mobile cellular technology is fast moving toward higher frequency bands and as there is a lot of scope for Multi-Input Multi-Output (MIMO) systems and Local Multi-Point Distribution System (LMDS) technology, it is expected that the techniques developed in Chapters 4, 5, 6, 7, and 8 can be applied there as well. Pointers on possible extensions of the work done are also given.

All the simulations in this book were written and tested in MATLAB® Version R2012b, and are bug free. For detailed product information on MATLAB®, please contact The MathWorks Inc, Corporate Headquarters, United States of America:

The MathWorks, Inc.
3 Apple Hill Drive
Natick, MA 01760-2098 USA
Tel: 508-647-7000
Fax: 508-647-7001
E-mail: info@mathworks.com
Web: www.mathworks.com

Acknowledgments

Words do not suffice to record my deep sense of gratitude to my guide, benefactor, and mentor Prof. M. Harisankar, without whose constant encouragement, vision, and wisdom, this work would not have been completed. He was always a constant source of inspiration, and an endless reservoir of energy, and held the torch of knowledge to show me the correct path whenever I faltered. I owe him very much for his patience and wisdom. Prof. M. Ramachandra Kaimal inspired me throughout the course of this research activity. He was instrumental in providing many valuable suggestions, which saved me many hours. I thank both of them for the time and energy they spared for me.

I thank Dr. Sakuntala S. Pillai for providing me many times with information on general topics as well as for helping me attend various conferences. I had several hours of technical discussions with Dr. P.P. Mohanlal, Deputy Director, VSSC Thiruvananthapuram, who, despite his very busy work schedule, was always ready to clear my technical queries. I express my gratitude to both of them.

My former colleague Dr. Suresh Kumaraswamy helped several times in getting me the necessary literature on request through e-mail. My old undergraduate classmate and friend of several years, Dr. G. Abhilash of NIT Kozhikode did share some of his vast expertise on several topics, which helped me a lot. I thank both of them for their help and friendship.

My sincere thanks to the publishers of this book, CRC Press, especially Dr. Gagandeep Singh, who showed keen interest in getting the manuscript ready for printing in an elegant manner and so quickly.

Finally, I wish to place on record the constant support I received from my family including my wife Sobha and daughters Anuja and Aabha during the entire course of the work.

And as always, I thank the God Almighty for giving me the strength to pursue this research amid my teaching assignments.

K.C. Raveendranathan
indran@ieee.org

List of Abbreviations

ACI	Adjacent Channel Interference
AMPS	Advanced Mobile Phone System
ANC	Adaptive Noise Cancellation
ANFF	Adaptive Neuro-Fuzzy Filter
ANFIS	Adaptive Network based Fuzzy Inference System
ANN	Artificial Neural Network
AWGN	Additive White Gaussian Noise
BAN	Body Area Networks
BER	Bit Error Rate
BPSK	Binary Phase Shift Keying
BTS	Base Switching Station
CCI	Co-Channel Interference
CCM	Channel Covariance Matrix
CDMA	Code Division Multiple Access
CDPD	Cellular Digital Packet Data
CIR	Channel Impulse Response
CMA	Constant Modulus Algorithm
CNFF	Compensatory Neuro-Fuzzy Filter
DAB	Digital Audio Broadcast
DCR	Digital Cellular Radio
DD	Decision Directed
DECT	Digital European Cordless Telephony
DFE	Decision Feedback Equalizer
DPSK	Differential Phase Shift Keying
DSP	Digital Signal Processor
DVB	Digital Video Broadcast
EKF	Extended Kalman Filter
ESN	Electronic Serial Number
FAF	Fuzzy Adaptive Filter
FBF	Fuzzy Basis Function
FCC	Forward Control Channel
FIR	Finite Impulse Response
FIS	Fuzzy Inference System
FLS	Fuzzy Logic System
FSE	Fractionally Spaced Equalizer
FSK	Frequency Shift Keying
FVC	Forward Voice Channel

GFSK	Gaussian Frequency Shift Keying
GMSK	Gaussian Minimum Shift Keying
GSM	Global System for Mobile
ICI	Inter-Channel Interference
IIR	Infinite Impulse Response
ISDN	Integrated Services Digital Network
ISI	Inter-Symbol Interference
LAN	Local Area Network
LE	Linear Equalizer
LMDS	Local Multipoint Distribution System
LMS	Least Mean Square
LOS	Line of Sight
LTI	Linear Time Invariant
LTV	Linear Time Variant
MANFIS	Modified Adaptive Network based Fuzzy Inference System
MAP	Maximum A Posteriori Probability
MF	Membership Function
MIMO	Multi-Input-Multi-Output
MIN	Mobile Identification Number
MLP	Multi-Level Perceptron
MLSE	Maximum Likelihood Sequence Estimator
MLVA	Maximum Likelihood Viterbi Algorithm
MMDS	Multichannel Multipoint Distribution Service
MMSE	Minimum Mean Square Error
MRAN	Minimum Resource Allocation Network
MSC	Mobile Switching Center
MTSO	Mobile Telephone Switching Office
NADC	North American Digital Cellular
NLOS	Non Line of Sight
NLTV	Nonlinear Time Variant
NNC	Nearest Neighbor Classifier
OFDM	Orthogonal Frequency Division Multiplexing
OLS	Orthogonal Least Squares
OQPSK	Offset Quadrature Phase Shift Keying
PA	Power Amplifier
PAN	Personal Area Networks
PDC	Personal Digital Cellular
PG	Pseudo Gaussian
PHS	Personal Handy-Phone System
PSD	Power Spectral Density
PSK	Phase Shift Keying
PSTN	Public Switched Telephone Network
QAM	Quadrature Amplitude Modulation
RBF	Radial Basis Function
RCC	Reverse Control Channel

RF	Radio Frequency
RLS	Recursive Least Squares
RVC	Reverse Voice Channel
SCM	Station Class Mark
SER	Symbol Error Rate
SINR	Signal-to-Interference Noise Ratio
SIR	Signal-to-Interference Ratio
SNR	Signal-to-Noise Ratio
SUI	Stanford University Interim
TDL	Tapped Delay Line
TE	Transversal Equalizer
TSK	Takagi–Sugeno–Kang
UWB	Ultra-Wide Band
WB	Wide Band

1

Introduction

1.1 Introduction

Although the real world is analog, digital communication systems have evolved over the years as they have several advantages in reality. There has been an astronomical growth in digital communication systems in the past few decades. The demand for this type of communications has also increased over the years. The telephone networks which were originally used to carry baseband analog voice now support ISDN, LAN and so on. Wireless networking has emerged as its own discipline over the past decade. There are other advancements like Digital Cellular Radio (DCR) which demand an efficient usage of the available radio spectrum.

Digital communication systems are generally more complicated to design and more expensive. However, they have the following advantages over analog communication systems:

- Some functions are too expensive or impractical to implement in the analog domain. Examples are nonlinear functions, linear phase filters and 2-D filters.

- Digital systems are insensitive to component tolerances, aging and temperature drift.

- Digital system behavior is more predictable (tractable) and repeatable.

- Digital systems are easily re-programmable.

- Last, their size is independent of values and allows a high integration level.

In order to facilitate digital data communication with bandwidth efficiency, we need to use adaptive equalizers. The topic under discussion in this book is channel equalization in the presence of Co-Channel Interference (CCI) and Adjacent Channel Interference (ACI) in mobile broadband communication channels in the presence of Additive White Gaussian Noise (AWGN). There have been many solutions to the problem of active noise cancellation in mobile indoor channels. Several investigators have studied various active noise cancellers in detail. The main objectives of this book are modeling of a mobile

1

broadband communication channel and designing of a Neuro-Fuzzy Adaptive Equalizer for it.

1.2   Need for Equalizers

In digital transmission, the problem of Inter-Symbol Interference (ISI) is most often mitigated by receiving-end equalization. The two important issues in the design and implementation of equalizers are its complexity and its training. The purpose and usage of a channel equalizer is illustrated in Figure 1.1.

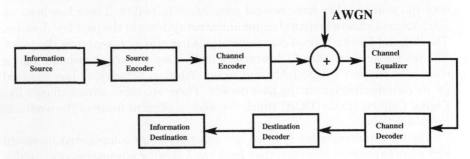

FIGURE 1.1
The Purpose of a Channel Equalizer in a Digital Communication System.

The development of the automatic linear adaptive equalizer in the late 1960s has paved the way for advance in digital communications. From this modest beginning adaptive equalizers have gone through several strides of development. The first generation equalizers are based on a linear adaptive filter algorithm with or without decision feedback. The Maximum Likelihood Sequence Estimator (MLSE) based on the Viterbi algorithm provides an alternative solution. These two solutions represent two extremes in adaptive equalizers—the linear equalizer is simple and computationally efficient, but it suffers from poor performance under extreme conditions. The infinite memory MLSE, on the other hand, provides good performance at the cost of large computational complexity. So there is need for a technique which is capable of achieving the best of the above two.

Now, advance in Digital Signal Processing provides scope for very large scale implementation of many complex algorithms in a lucid manner. The programming capability of Digital Signal Processors (DSPs) makes them very suitable for fast realization. This is definitely an advantage while developing a new system based on an unproven technology, as modifying the design is a matter of re-programming the DSP chip. Thus the product development cycle gets simplified. And modifications can be incorporated quite easily.

Due to the reasons mentioned above, nonlinear equalizers are being investigated by many researchers. They include Artificial Neural Networks (ANN), Radial Basis Functions (RBF), recurrent networks and neuro-fuzzy systems. In this work, we implement such a nonlinear, fuzzy adaptive equalizer for use in mobile broadband and Ultra-Wide Band (UWB) communication channels.

1.3   Review of Contemporary Literature

As the topic of research comprises two disciplines—equalization problems and neuro-fuzzy control systems—there are a number of reference sources for each one of them. A detailed discussion on the LMS algorithm which paved the way to the development of linear equalizers, can be found in (Widro 1975 and Widro and Hoff 1960). But the first practical implementation of an adaptive channel equalizer was done by Robert W. Lucky in 1965 (Lucky 1965). It was soon found out that, though they were simple to implement, they were unsuitable for highly dispersive channels. Soon Forney had implemented the Viterbi Algorithm for the MLSE equalizer (Forney 1978). But this suffered from computational complexity. At the same time, IIR implementations of linear equalizers also came, which, with feedback, could get better performance and were called Decision Feedback Equalizers (DFE). Subsequently, computationally efficient algorithms like the Recursive Least Square (RLS) algorithm, Kalman filters and the RLS lattice algorithm were also developed.

Developments in the field of Artificial Neural Networks (ANN), a nonlinear signal processing methodology, took place in the late 1980s. The Multi-Layer Perceptron (MLP) was developed in 1990. Another nonlinear processor was the RBF. The seminal paper by Lotfi A. Zadeh (Zadeh 1965) in 1965 had opened up a new way in the thinking of the design of logic systems altogether. In a subsequent paper (Zadeh 1973) he showed how to apply *fuzzy logic* to typical control system paradigms. It was Bart Kosko (Kosko 1990) who first thought about unsupervised learning in the presence of noise. Later several authors tried to combine the fuzzy logic principles with artificial neural networks, and a new control paradigm, viz., the *neuro-fuzzy system*, was born. In relatively recent times, many researchers have successfully developed channel equalizers using fuzzy logic (Sarwal and Srinath 1995, Komninakis et al. 2000, Lee 1996, Li et al. 2003, Patra and Mulgrew 1998). Another recent trend is to make use of *type-2 Fuzzy Adaptive Filters* to design channel equalizers for time varying cellular mobile channels (Chen et al. 1995, Liang and Mendel 2000).

The mobile cellular channel is known to be a *Linear Time Variant (LTV)* channel in general. It is also known that it is either a *Rayleigh fading* or *Ricean fading* channel, depending on the number of modes that reach the receiver through multiple paths. The fading characteristics will be those of a

Ricean distribution, if apart from the major ray, one more component reaches the receiver *(two-ray model)*. It will exhibit a *Rayleigh distribution* if three or more multipath components reach the receiver. Typically, mobile channels are severely affected by CCI due to *frequency re-use* and ACI, due to the leakage of spectrum (due to imperfect receiver filtering) from adjacent channels allocated within a cell. They are also affected by noise, which is normally modeled as AWGN. Several models are available for mobile cellular channels, thereby characterizing the *Channel Impulse Response (CIR)*. The Inter-Channel Interference (ICI) and ACI contribute to reducing the output Signal-to-Noi: (SNR). Active Noise Cancellation (ANC) uses artificial signals to cancel undesired noise. A modified *fuzzy adaptive filtered-X algorithm* was considered by Chang (Chang and Shyu 2003). The modified *fuzzy adaptive filtered-X algorithm* can be applied to a mobile cellular channel in the indoor environment. Adaptive noise cancellation using Fuzzy Neural Networks has recently come into limelight again (Meng 2005). Space-Time Equalization for mobile broadband communication in an industrial indoor environment is discussed in Trautwein et al.1999.

It is established that *Blind Equalization* is more suitable for broadcast channels like the mobile cellular channels.[1] Blind equalization is applied to eliminate the channel distortion and multipath effects since the transmitted signal is unknown at the receiver end. The purpose of blind channel equalization is to remove ISI caused by time dispersion in the channel response without resorting to an explicit knowledge of the channel characteristics or the channel input sequence (Dogancay and Kennedy 1999). The interest in *blind equalization using RBF neural networks* has been revived by Nan Xie (Xie and Leung 2005). Recently some authors have contributed considerably in the construction of equalizers for broadcast channels. Blind equalization using Pseudo-Gaussian based Compensatory Neuro-Fuzzy Filter (CNFF) was one such approach (Lin and Ho 2003, Lin and Juang 1996). We can adapt the CNFF for the indoor mobile cellular channel. Lambert et al. (1996) described an adaptive block decision feedback receiver for improved performance in channels with severe ISI. The theory of an Adaptive Network-based Neuro-Fuzzy Inference System (ANFIS) and its application in nonlinear problem solving was first suggested by Jang in his seminal paper (Jang 1993). Several channel models are discussed in (IEEE 802.16 BWA WG 2000). This led to several new strides in system identification and control system design. Design and simulation of mobile channel equalizers based on ANFIS is one of the major areas of focus of this book.

---

[1] In blind equalizers, there is no transmission of a training sequence to adaptively adjust the equalizer network prior to actual transmission of information.

1.4 Major Contributions of the Book

The major contributions of this book are summarized in this section. The mobile cellular channel can be modeled as an *LTV* channel in general. Several specific channel models exist in mobile communication scenarios, which are chosen according to the particular terrain conditions. It is known that the most suitable model for a mobile radio channel in an indoor environment is that of a multipath *Rayleigh fading* model. The above fact is established based on earlier works.

- The concept of a type-2 Fuzzy Adaptive Filter (FAF-II)-based equalizer for LTV channels is given and its performance is compared with earlier reported FAF-I and Neural Network Classifier (NNC) for the same purpose. It is concluded that FAF-II outperformed FAF-I and NNC.

- An equalizer based on ANFIS is implemented for the aforesaid channel. It is also shown that ANFIS-based equalizers can be adapted for UWB channels as well.

- A CNFF is considered for the above channel to combat CCI. A detailed performance analysis is made.

- A generic framework based on a radial basis function neural network, for the above three implementations of equalizers, is established. This framework is most useful in comparing the performances of FAF-II, ANFIS, and CNFF equalizers.

- The modeling of a mobile cellular channel as a nonlinear time variant (NLTV) system has not been attempted so far by any investigators. A novel modular approach toward the simulation and modeling of NLTV channels is considered. It is shown that this approach is quite suitable for situations where the modulator/Power Amplifier (PA) stages are driven to nonlinear regions (to increase power efficiency).

Further Reading

B. Widrow et al., The Complex LMS Algorithm, *Proceedings of the IEEE*, Vol.63, pp.719–720, April 1975.

B. Widrow and M.E. Hoff, Adaptive Switching Circuits, *Proceedings of the IRE WESCON conv.*, pp.94–104, August 1960.

Bart Kosko, Unsupervised Learning in Noise, *IEEE Transactions on Neural Networks*, Vol.1, No.1, pp.44–57, March 1990.

C.Y. Chang and K.K. Shyu, Active Noise Cancellation with a Fuzzy Adaptive Filtered-X Algorithm, *IEE Proceedings—Circuits Devices Systems*, Vol.150, No.5, pp.416–422, October 2003.

Cheng Jian Lin and Wen Hao Ho, Blind Equalization Using Pseudo-Gaussian-Based Compensatory Neuro-Fuzzy Filters, *International Journal of Applied Science and Engineering*, Vol.2, No.1, pp.72–89, January 2004.

Cheng Jian Lin and Wen Hao Ho, A Pseudo-Gaussian-Based Compensatory Neuro-Fuzzy System, *Proceedings of the 12th IEEE International Conference on Fuzzy systems, FUZZ-03*, Vol.1, pp.214–219, May 2003.

Chin Teng Lin, and Chia Feng Juang, An Adaptive Neural Fuzzy Filter and Its Applications, *Proceedings of the Fifth IEEE International Conference on Fuzzy Systems*, Vol.1, pp.564–569, September 1996.

Christos Komninakis et al., Adaptive Multi-Input Multi-Output Fading Channel Equalization Using Kalman Estimation, *Proceedings of the ICC 2000*, Vol.3, pp.1655–1659, June 2000.

Davis E. Lambert, N.A. Pendergrass, and Susan M. Jarvis, An Adaptive Lock Decision Feedback Receiver for Improved Performance in Channels with Severe Intersymbol Interference, *Proceedings of the MTS/IEEE Conference on Prospects for the 21st Century, OCEANS-96*, Vol.2, pp.984–987, September 1996.

G.D. Forney, Maximum-Likelihood Sequence Estimation of Digital Sequences in the Presence of Intersymbol Interference, *IEEE Transactions on Information Theory*, Vol.IT-18, pp.363–378, May 1978.

IEEE Working Group, Channel Models for Broadband Fixed Wireless Systems, *IEEE 802.16 Broadband Wireless Access Working Group Document*, 2000.

Jyh-Shing Roger Jang, ANFIS: Adaptive-Network-Based Fuzzy Inference System, *IEEE Transactions on Systems, Man, and Cybernetics*, Vol.23, No.3, pp.665–685, May/June 1993.

Ki Yong Lee, Complex Fuzzy Adaptive Filter with LMS Algorithm, *IEEE Transactions on Signal Processing*, Vol.44, No.2, pp.424–427, February 1996.

Kutluyil Dogancay and Rodney A. Kennedy, Least Squares Approach to Blind Channel Equalization, *IEEE Transactions on Communications*, Vol.47, No.11, pp.1678–1687, November 1999.

Lotfi A. Zadeh, Fuzzy Sets, *Information and Control*, Vol.8, pp.338–353, 1965.

Lotfi A. Zadeh, Outline of a New Approach to the Analysis of Complex Systems and Decision Processes, *IEEE Transactions on Systems, Man and Cybernetics*, Vol.SMC-3, No.2, pp.28–44, January 1973.

Meng Joo Er, Zhengrong Li, Huaning Cai, and Qing Chen, Adaptive Noise Cancellation Using Enhanced Dynamic Fuzzy Neural Networks, *IEEE Transactions on Fuzzy Systems*, Vol.13, No.3, pp.331–342, June 2005.

Nan Xie and Henry Leung, Blind Equalization Using a Predictive Radial Basis Function Neural Network, *IEEE Transactions on Neural Networks*, Vol.16, No.3, pp.709–720, May 2005.

P. Sarwal and M.D. Srinath, A Fuzzy Logic System for Channel Equalization, *IEEE Transactions on Fuzzy Systems*, Vol.3, No.2, pp.246–249, May 1995.

Qilian Liang and Jerry M. Mendel, Equalization of Nonlinear Time-Varying Channels Using Type-2 Fuzzy Adaptive Filters, *IEEE Transactions on Fuzzy Systems*, Vol.8, No.5, pp.551–563, October 2000.

Robert W. Lucky, Automatic Equalization of Digital Communication, *Bell System Tech. Journal*, Vol.44, pp.547–588, April 1965.

Sarat Kumar Patra and B. Mulgrew, Efficient Architecture for Bayesian Equalization Using Fuzzy Filters, *IEEE Transactions on Circuits and Systems–II: Analog and Digital Signal Processing*, Vol.45, No.7, pp.812–820, July 1998.

Sheng Chen et al., Adaptive Bayesian Decision Feedback Equalizer for Dispersive Mobile Radio Channels, *IEEE Transactions on Communications*, Vol.43, No.5, pp.1937–1945, May 1995.

U. Trautwein et al., A Simulation Study on Space-Time Equalization for Mobile Broadband Communication in an Industrial Indoor Environment, *Proceedings of the IEEE Vehicular Technology Conference*, May 16–20, 1999.

Zhengrong Li, Meng Joo Er, and Yang Gao, An Adaptive RBFN based Filter for Adaptive Noise Cancellation, *Proceedings of the 42nd IEEE Conference on Decision and Control*, Mani, HI, USA, pp.6175–6180, December 2003.

2

Overview of Mobile Channels and Equalizers

2.1 Introduction

This book discusses the analysis of channel equalizers for mobile cellular channels in the presence of a number of linear and nonlinear message corrupting mechanisms. To understand properly the context and need of the work, a detailed discussion of the fundamental concepts involved is essential. This chapter focuses on the need for equalizers in a mobile channel and introduces different models available for the mobile channel. Various currently available equalizers are also discussed.

The chapter is organized as follows. In Section 2.2 the mobile communication system is reviewed and Section 2.3 brings out fading characteristics of mobile channels. In Section 2.4, various models available for mobile cellular channels are discussed. In Section 2.5, a classification of equalizers is done. Section 2.6 provides concluding remarks.

2.2 Mobile Cellular Communication System

In mobile cellular radio a large number of low-power base stations for transmission are deployed to cover a limited area. A cellular telephone system provides a wireless connection to a Public Switched Telephone Network (PSTN) for any user location within the radio range of the system. Cellular systems accommodate a large number of users over a large geographic area, within a limited frequency spectrum, with comparable quality of service to that of the landline telephone system (T.S.Rappaport 2003). High capacity is achieved by limiting the coverage of each base station transmitter to a small geographic area called a *cell*, so the same radio channels may be reused by another base station located at a distance. A sophisticated switching technique called a *handoff* enables a call to proceed uninterrupted when a user moves from one cell to another.

Figure 2.1 shows a basic cellular system which consists of *mobile stations, base stations and a mobile switching center (MSC)*. The MSC/*mobile telephone switching office (MTSO)* connects all mobiles to the PSTN. The base stations/*Base Switching Stations (BTS)* are connected to the MTSO by microwave/Optical Fiber Cable (OFC) links. The MTSO coordinates the activities of all of the base stations and connects the entire cellular system to the PSTN. A typical MTSO handles 100,000 cellular subscribers and 5,000 simultaneous conversations at a time, and accommodates all billing and systems maintenance functions as well (T.S.Rappaport 2003). The channels used for

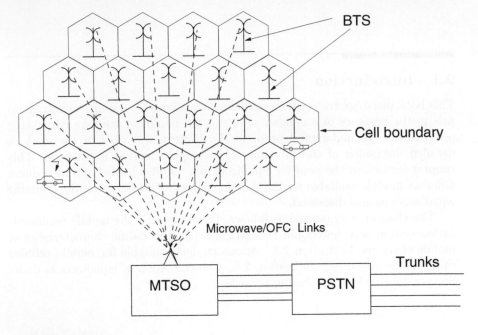

FIGURE 2.1
The Organization of a Mobile Cellular System.

voice transmission from BTS to mobiles are called *Forward Voice Channels (FVC)*, and the channels used for voice transmissions from mobiles to the BTS are called *Reverse Voice Channels (RVC)*. The two channels responsible for initiating mobile calls are the *Forward Control Channels (FCC)* and *Reverse Control Channels (RCC)* (Rappaport 2003).

2.2.0.1   Call Initiation

When a cell phone is turned on, but not yet engaged in a call, it first scans the group of FCCs to determine the one with the strongest signal, and then monitors that control channel until the signal drops below a usable level. At this point, it again scans the control channels in search of the strongest BTS signal. When a telephone call is placed to a mobile user, the MTSO dispatches the request to all the base stations in the cellular system. The *Mobile Identification Number (MIN)* (the subscriber's telephone number) is then broadcast as a paging message over all of the FCCs throughout the cellular system. The mobile receives the paging message sent by the BTS which it monitors, and responds by identifying itself over the RCC. The BTS relays the acknowledgment sent by the mobile and informs the MTSO of the handshake. Then the MSC instructs the BTS to move the call to an unused voice channel within the cell (typically, between ten to sixty voice channels and just one control channel are used in each cell's base station). Now the base station signals the mobile to change frequencies to an unused FVC and RVC pair, at which point another data message (alert) is transmitted over the FVC to instruct the cell phone to ring, thereby instructing the mobile user to answer the phone (Rappaport 2003). Once a call is in progress, the MSC adjusts the transmitted power of the mobile (cell phone) and changes the channel of the cell phone and base stations in order to maintain call quality as the subscriber moves in and out of range of each base station. This is called *handoff*.

When a mobile originates a call, a call initiation request is sent on the RCC. With this request the mobile unit transmits its telephone number (MIN), *Electronic Serial Number (ESN)*, and the telephone number of the called party. The mobile also transmits a *Station Class Mark (SCM)* which indicates what the maximum transmitter power level is for the particular user. The cell BTS receives this data and sends it to the MTSO. The MTSO validates the request, makes a connection to the called party through the PSTN (if necessary), and instructs the BTS and mobile user to move to an unused forward and reverse voice channel pair to allow the conversation to begin (Rappaport 2003).

2.2.0.2 Frequency Reuse

Mobile cellular systems rely on an intelligent allocation and reuse of channels throughout a coverage region. Each cellular base station is allocated a group of radio channels to be used within a small geographic area called a *cell*. Consider a cellular system which has a total of S duplex channels available for use. If each cell is allocated a group of k channels ($k < S$), and if the S channels are divided among N cells into unique and disjoint channel groups which each have the same number of channels, the total number of available radio channels is given by $S = kN$. The N cells which collectively use the complete set of available frequencies are called a *cluster*. If a cluster is replicated M times within the system, the total number of duplex channels, C, can be used as a measure of capacity and is given by $C = MkN = MS$. Base stations in adjacent cells are assigned channel groups which contain completely different channels than the neighboring cells (Rappaport 2003). The base station antennas are designed to achieve the desired coverage within the particular cell. By limiting coverage to within the boundaries of a cell, the same group of channels may be used to cover different cells that are separated from one another by distances large enough to keep interference levels within tolerable limits. This concept is called *frequency reuse/frequency planning*. Frequency reuse implies that in a given coverage area there are several cells that use the same set of frequencies. By using the hexagon geometry for the cells, the fewest number of them can cover a geographic region. Moreover, the hexagon closely approximates the circular radiation pattern for an omni-directional base station antenna (Rappaport 2003). It can be shown that when the cluster size N is 7, the frequency reuse factor is $1/7$. In order to connect without gaps between adjacent cells, the geometry of the hexagon is such that the number of cells per cluster, N, can only have values which satisfy Equation 2.1, where i and j are nonnegative numbers.

$$N = i^2 + ij + j^2 \tag{2.1}$$

To find the nearest co-channel neighbors of a particular cell, one must move i cells along any chain of hexagons and then turn 60 degrees counter-clockwise and move j cells (Rappaport 2003).

2.2.1   Co-Channel Interference and System Capacity

With frequency reuse, there are several cells that use the same frequency, which are called *co-channels*. The interference between signals from these cells is called *Co-Channel Interference (CCI)*. To reduce CCI, co-channel cells must be physically separated by a minimum distance to provide sufficient isolation due to propagation. When the size of each cell is nearly the same and the base stations transmit the same power, the CCI ratio is independent of the transmitted power and becomes a function of the radius of the cell (R) and the distance between centers of the nearest co-channel cells (D). By increasing the D/R ratio, the spatial separation between co-channel cells in relation to the coverage distance of a cell is increased. Thus CCI is reduced, due to the improved isolation of RF energy from the co-channel cell. The parameter Q, called the *co-channel reuse ratio,* is related to the cluster size. For a hexagonal geometry, $Q = D/R = \sqrt{3N}$. A small value of Q provides larger capacity since the cluster size N is small, whereas a large value of Q improves the transmission quality, due to a smaller value of CCI. Hence, a trade-off between these two must be made in actual design (Rappaport 2003). If i_0 is the number of interfering co-channels, then the signal–to–interference ratio (SIR) for a mobile receiver can be expressed as

$$SIR = \frac{S}{\sum\limits_{i=0}^{i_0} I_i} \tag{2.2}$$

Propagation measurements show that in a mobile radio channel, the average received power P_r at a distance d from the transmitting antenna is approximated by the formula

$$P_r = P_0 \left(\frac{d}{d_0} \right)^{-n} \tag{2.3}$$

where P_0 is the power received at a close-in reference point in the far field region of the antenna, at a small distance d_0 from the transmitting antenna, and n is the path loss exponent. The value of n typically ranges from 2 to 4 in urban cellular systems (Rappaport 1992). If D_i is the distance of the ith interferer from the mobile, the received power at a given mobile due to the ith interfering cell will be proportional to $(D_i)^{-n}$. When the transmit power of each base station is equal and the path loss exponent is the same throughout the coverage area, the SIR for a mobile can be approximated as

$$SIR = \frac{R^{-n}}{\sum_{i=1}^{i_0} (D_i)^{-n}} \tag{2.4}$$

Considering only the first layer of interfering cells, if all the interfering base stations are equidistant from the desired base station and if this distance is equal to the distance D between cell centers, then Equation 2.4 simplifies to

$$SIR = \frac{(D/R)^n}{i_0} = \frac{(\sqrt{3N})^n}{i_0} = \frac{Q^n}{i_0} \tag{2.5}$$

Using an exact cell geometry layout, it can be shown for a seven-cell cluster (N = 7), with the mobile cell unit at the cell boundary, the mobile is at a distance $D - R$ from the two nearest co-channel interfering cells and is exactly $D + R/2$, D, $D - R/2$, and $D + R$ from the other interfering cells in the first tier (Lee 1986). Therefore Equation 2.4 can be written approximately as (assuming the path loss exponent, $n = 4$)

$$SIR = \frac{1}{2 \times [(Q-1)^{-4} + (Q+1)^{-4} + Q^{-4}]} \tag{2.6}$$

For $N = 7$, the co-channel reuse ratio, $Q = \frac{D}{R} = \sqrt{3N} = 4.6$, and the worst case SIR is exactly 49.56 (16.95 dB) (Rappaport 2003). A plot of N versus SIR is shown in Figure 2.2. The MATLAB code to generate the same is appended below.

```
%%% MATLAB code to plot Cluster Size versus SIR..
clear all; close all; clf;
n=[];
for i=1:20
    for j=1:20
        n=[n,i^2+i*j+j^2];
    end;
end;
N=sort(n);
Q=sqrt(3*N);
sir=1./(2*((Q-1).^(-4)+(Q+1).^(-4)+Q.^(-4)));
sirdb=10*log10(sir);
plot(N,sirdb,'LineWidth', 2);
xlabel('Cluster Size, N');
```

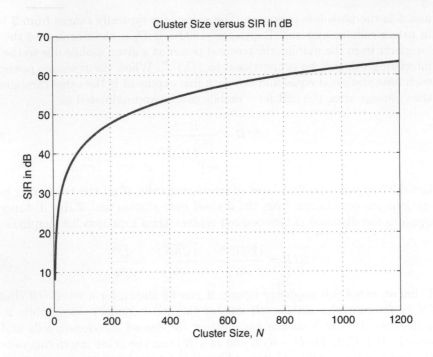

FIGURE 2.2 (See color insert.)
A Plot of Cluster Size versus SIR in dB.

```
ylabel('SIR in dBs');
title('Cluster Size versus SIR in dBs');
grid;
%%% end of nsir.m..
```

For the US Advanced Mobile Phone System (AMPS) cellular system which uses FM and 30 MHz channels, subjective tests indicate that sufficient voice quality is provided when $SIR \geq 18$ dB. Thus to meet this condition, it would be necessary to increase N to the next larger size, which from Equation 2.1 is found to be 12 (corresponding to $i = j = 2$). This obviously results in a significant decrease in capacity, since 12-cell reuse offers a spectrum utilization of $1/12$ within each cell. From the above discussion, *it is clear that CCI determines link performance, which in turn dictates the frequency reuse plan and the overall capacity of cellular systems* (Rappaport 2003).

2.2.2 Adjacent Channel Interference

Interference from signals which are adjacent in frequency to the desired signal is called *Adjacent Channel Interference* (ACI). ACI results from imperfect receiver filters which allow nearby frequencies to leak into the passband. The problem can be particularly serious if an adjacent channel user is transmitting in very close range to a subscriber's receiver, while the receiver attempts to receive a base station on the desired channel. This is referred to as the *near–far* effect, where a nearby transmitter (which may or may not be of the same type as that used by the cellular system) captures the receiver of the subscriber. Also when a mobile close to a base station transmits on a channel close to one being used by a weak mobile, the near–far effect occurs (Rappaport 2003).

ACI can be minimized through careful filtering and channel assignments. Since each cell is given only a fraction of the available channels, a cell need not be assigned channels which are adjacent in frequency. By keeping the frequency separation between each channel in a given cell as large as possible, the ACI may be reduced considerably. Thus instead of assigning channels which form a contiguous band of frequencies within a particular cell, channels are allocated in such a way that the frequency separation between channels is maximized. There exist many channel allocation schemes that are able to separate adjacent channels in a cell by as many as N channel bandwidths, where N is the cluster size (Rappaport 2003).

If the frequency reuse factor is large (e.g., small N), the separation between adjacent channels at the base station may not be sufficient to keep the adjacent channel interference level within tolerable limits. For example, if a close-in mobile is 20 times as close to the base station as another mobile and has energy spill out of its passband, the SIR at the base station for the weak mobile (before receiver filtering) is approximately

$$\frac{S}{I} = (20)^{-n} \tag{2.7}$$

where n is the path loss exponent. For $n = 4$, SIR is -52 dB (Rappaport 2003).

2.2.3 Digital Modulation Types and Relative Efficiencies

This section covers the main digital modulation formats and their variants used in practical systems, as applicable to cellular telephony, along with their relative spectral efficiencies. Fortunately, there are a limited number of modulation types, which form the building blocks of any system. Table 2.1 covers the applications of different modulation formats in both wireless communications and video (Peterson et al. 1995). Bandwidth efficiency describes how efficiently the allocated bandwidth is utilized or the ability of a modulation scheme to accommodate data, within a limited bandwidth. Note that the figures given in Table 2.1 for *theoretical bandwidth efficiency* cannot actually be achieved in practical radios, since they require perfect modulators, demodulators, filters and transmission paths. If the radio has a perfect (rectangular in frequency domain) filter, then the occupied bandwidth can be made equal to the symbol rate. Techniques for maximizing spectral efficiency include the following:

TABLE 2.1
Digital Modulation Formats

Mod. Format	Theo. B.W. η	Applications
MSK, GMSK	1bps/Hz	GSM, CDPD
BPSK	1bps/Hz	Deep Space Telemetry, Cable Modems
QPSK, DQPSK	2bps/Hz	Satellite, CDMA, LMDS,DVB-S, Cable Modems
OQPSK	2bps/Hz	CDMA, Satellite.
FSK, GFSK	0.4bps/Hz	DECT paging, AMPS, CT2, Land Mobile
8PSK	3bps/Hz	Satellite, Aircraft, Telemetry Pilot
16QAM	4bps/Hz	μ−wave Digital Radio, Modems, DVB-C/T
32QAM	5bps/Hz	Terrestrial Microwave,DVB-T
64QAM	6bps/Hz	DVB-C, Modems, Set Top Boxes, MMDS
256QAM	8bps/Hz	Modems, DVB-C, Digital Video

1. Relate the data rate to the frequency shift (as in GSM).

2. Use pre-modulation filtering to reduce the occupied bandwidth. Raised cosine filters as used in North American Digital Cellular (NADC), Personal Digital Cellular (PDC)—a 2G system—, and Personal Handy-Phone System (PHS), give the best spectral efficiency.

3. Restrict the types of transitions.

2.3   Fading Characteristics of Mobile Channels

In mobile cellular radio transmission between a base station and a mobile telephone, the signal transmitted from the base station to the mobile receiver is usually reflected from surrounding buildings, hills, and other obstructions. As a consequence, we observe multiple propagation paths arriving at the receiver at different delays. Hence the received signal has characteristics similar to those for ionospheric propagation. The same is true for transmission from the mobile telephone to the base station. Moreover, the speed that the mobile (automobile, train, etc.) is traveling results in frequency offsets, called *Doppler shifts*, of the various frequency components of the signal (Proakis and Salehi

2002). As the intervening medium changes its characteristics with respect to time, the mobile radio channel is time varying.

2.3.0.1 Tapped Delay Line (TDL) Channel Model

A general model for a time-variant multipath channel is a TDL structure. It consists of a tapped delay line with uniformly spaced taps. The tap spacing between adjacent taps is $1/W$, where W is the bandwidth of the signal transmitted through the channel. The tap coefficients, denoted as $\{c_n(t) \equiv \alpha_n(t)\, e^{j\phi_n(t)}\}$, are usually modeled as complex valued, Gaussian random processes which are mutually uncorrelated (Proakis and Salehi 2002).

2.3.0.2 Rayleigh and Ricean Fading Models

We can express each of the tap coefficients as

$$c_n(t) = c_r(t) + jc_i(t) \tag{2.8}$$

where $c_r(t)$ and $c_i(t)$ represent real-valued Gaussian random processes. We assume that $c_r(t)$ and $c_i(t)$ are stationary and statistically independent. We can also express $c_n(t)$ as

$$c_n(t) \equiv \alpha_n(t)\, e^{j\phi_n(t)} \tag{2.9}$$

where

$$
\begin{aligned}
\alpha_n(t) &= \sqrt{c_r^2(t) + c_i^2(t)} \\
\phi_n(t) &= tan^{-1}\frac{c_i(t)}{c_r(t)}
\end{aligned}
\tag{2.10}
$$

Now, if $c_r(t)$ and $c_i(t)$ are Gaussian with zero mean values, the amplitude of $\alpha_n(t)$ is characterized by the Rayleigh probability density function and $\phi_n(t)$ is uniformly distributed over the interval $(0, 2\pi)$. As a consequence the channel is called a *Rayleigh fading channel*. The Rayleigh fading signal amplitude is given by the probability density function as

$$f(\alpha) = \frac{\alpha}{\sigma^2}\, e^{-\alpha^2/2\sigma^2}, \quad \alpha \geq 0 \tag{2.11}$$

and $f(\alpha) = 0$ for $\alpha < 0$. The parameter $\sigma^2 = E(c_r^2) = E(c_i^2)$.

On the other hand, if $c_r(t)$ and $c_i(t)$ are Gaussian with nonzero mean values, the amplitude of $\alpha_n(t)$ is characterized by the Rice probability density function and $\phi_n(t)$ is also nonzero mean. In this case the channel is called a *Ricean fading channel* and the probability density function of the amplitude is given as

$$f(\alpha) = \frac{\alpha}{\sigma^2} e^{-(\alpha^2+s^2)/2\sigma^2} I_o\left(\frac{s\alpha}{\sigma^2}\right), \quad \alpha \geq 0 \tag{2.12}$$

where the parameter s^2 represents the power of the received nonfading signal component and $\sigma^2 = VAR(c_r) = VAR(c_i)$ (Proakis and Salehi 2002).

2.4   Channel Models

An important requirement for assessing technology for mobile radio applications is to have an accurate description of the wireless channel. Channel models are heavily dependent upon the radio architecture. For example, in the first generation systems, a super-cell or "single-stick" architecture is used where the BTS and the subscriber station are in Line-Of-Sight (LOS) condition and the system uses a single cell with no co-channel interference. For second generation systems, a scalable multicell architecture with Non-Line-Of-Sight (NLOS) conditions becomes necessary. Typically, the scenario is as follows:

- Cells are < 10 km in radius, variety of terrain and tree density types

- Under-the-eave/window or rooftop installed directional antennas (2–10 m) at the receiver

- 15–40 m BTS antennae

- High cell coverage requirement (80–90%)

The wireless channel is characterized by:

- Path loss (including shadowing)

- Multipath delay spread

- Fading characteristics

- Doppler spread

- Co-channel and adjacent channel interference

It is to be noted that these parameters are random and only a statistical characterization is possible (IEEE 802.16 BWA WG 2000). Typically, the mean and variance of parameters are specified. The above propagation model parameters depend upon terrain, tree density, antenna height, antenna beamwidth, wind speed, season (time of the year).

2.4.1   Suburban Path Loss Model

The most widely used path loss model for signal strength prediction and simulation in macrocellular environments is the *Hata–Okumura model* (Okumura 1968, Hata 1980). This model is valid for the 500–1500 MHz frequency range, receiver distances greater than 1 km from the base station, and base station antenna heights greater than 30 m. There exists an elaboration on the Hata–Okumura model that extends the frequency range up to 2000 MHz. It

TABLE 2.2

Channel Parameters for Suburban Path Loss Models

Model Parameter	Terrain Type A	Terrain Type B	Terrain Type C
a	4.6	4	3.6
b	0.0075	0.0065	0.005
c	12.6	17.1	20

was found that these models are not suitable for lower base station antenna heights, and hilly or moderate-to-heavy wooded terrain. However, other subcategories and different terrain types can be found around the world.

The maximum path loss category is hilly terrain with moderate-to-heavy tree densities (Category A). The minimum path loss category is mostly flat terrain with light tree densities (Category C). The intermediate path loss condition is captured in Category B. The extensive experimental data was collected by AT&T Wireless Services across the United States in 95 existing macrocells at 1.9 GHz. For a given close-in distance d_0, the median path loss (PL in dB) is given by

$$PL = A + 10\gamma \log_{10}(d/d_0) + s \quad for \quad d > d_0, \tag{2.13}$$

where $A = 20\log_{10}(4\pi d_0/\lambda)$, $l(\lambda$ being the wavelength in m), γ is the path loss exponent with $\gamma = (a - b\,h_b + c/h_b)$ for h_b between 10 and 80 m (h_b is the height of the base station in m), $d_0 =$100 m and a, b, c are constants dependent on the terrain category given in Table 2.2. The shadowing effect is represented by s, which follows log-normal distribution. The typical value of the standard deviation for s is between 8.2 and 10.6 dB, depending on the terrain/tree density type.

2.4.2   Urban (Alternative Flat Suburban) Path Loss Model

It has been shown that the Cost 231 Walfish-Ikegami (W-I) model (Smith and Dalley 2000)   matches extensive experimental data for flat suburban and urban areas with uniform building height. It has also been found that the model presented in the previous section for Category C (flat terrain, light tree density) is in a good agreement with the Cost 231 W-I model for suburban areas, providing continuity between the two proposed models. Figure 2.3 compares a number of published path loss models for suburban morphology with an empirical model based on drive tests in the Dallas–Fort Worth area in the United States.   The Cost 231 Walfish-Ikegami model is used with the following parameter settings: Frequency = 1.9 GHz, Mobile Height = 2 m, Base Height = 30 m, Building spacing = 50 m, Street width = 30 m, Street orientation = 90°. It has also been found that the Cost 231 W-I model agrees well with measured results for urban areas, provided the appropriate building spacing

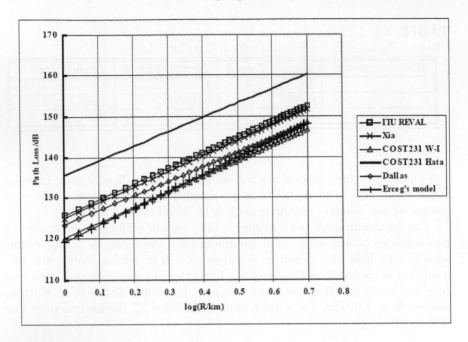

FIGURE 2.3
Comparison of Different Suburban Path Loss Models: Path Loss in dB versus
Logarithm of Distance in Kilometers.

and rooftop heights are used. It can therefore be used for both suburban and
urban areas, and can allow for variations of these general categories between
and within different countries. Flat terrain models in conjunction with terrain
diffraction modeling for hilly areas can be used in computer based propagation tools that use digital terrain databases. In Smith and Dalley (2000), it
is shown that the weighting term for knife-edge diffraction should be set to
0.5 to minimize the log-normal standard deviation of the path loss.

2.4.2.1 Multipath Delay Profile

Due to the scattering environment, the channel has a multipath delay profile.
For directive antennas, the delay profile can be represented by a spike-plus-
exponential shape. It is characterized by τ_{rms} (RMS delay spread of the entire
delay profile) which is defined as

$$\tau_{rms}^2 = \sum_j P_j \tau_j^2 - (\tau_{avg})^2 \qquad (2.14)$$

where $\tau_{avg} = \sum_j P_j \tau_j$, τ_j is the delay of the jth delay component of the profile and P_j is given by

$$P_j = \frac{power \ in \ the \ j^{th} \ delay \ component}{total \ power \ in \ all \ components}.$$

The delay profile has been modeled using a spike-plus-exponential shape given by

$$P(\tau) = A\delta(\tau) + B \sum_{i=0}^{\infty} \exp(-i\Delta\tau/\tau_0)\delta(\tau - i\Delta\tau), \tag{2.15}$$

where A, B and $\Delta\tau$ are experimentally determined.

2.4.2.2 RMS Delay Spread

A delay spread model is proposed in Greenstein et al. (1997) based on a large body of published reports. It is found that the RMS delay spread follows log-normal distribution and that the median of this distribution grows as some power of distance. The model is developed for rural, suburban, urban, and mountainous environments. The model is of the following form.

$$\tau_{rms} = T_1 d^\varepsilon y \tag{2.16}$$

where τ_{rms} is the RMS delay spread, d is the distance in km, T_1 is the median value of τ_{rms} at $d = 1$ km, ε is an exponent that lies between 0.5–1.0, and y is a log-normal variate. The model parameters are valid only for omni-directional antennas. It is shown that 32° and 10° directive antennas reduce the median τ_{rms} values for omni-directional antennas by factors of 2.3 and 2.6, respectively. Depending on the terrain, distances, antenna directivity, and other factors, the RMS delay spread values can vary from very small values (tens of nanoseconds) to large values (microseconds).

2.4.2.3 Fade Distribution, K-Factor

Narrow band received signal fading can be characterized by a Ricean distribution. The key parameter of this distribution is the K-factor, defined as the ratio of the *fixed component power* and the *scatter component power*. The narrowband K-factor distribution is found to be log-normal, with the median as a simple function of season, antenna height, antenna beamwidth, and distance. The standard deviation is found to be approximately 8 dB. The model presented in Greenstein et al. (1999) is as follows.

$$K = F_s F_h F_b K_o d^\gamma u \tag{2.17}$$

where F_s is a seasonal factor, F_s =1.0 in summer (leaves) and 2.5 in winter (no leaves), F_h is the receive antenna height factor, $F_h = $ (h/3)0.46 (h is the receive antenna height in meters), F_b is the beamwidth factor, $F_b = $ (b/17)-0.62; (b in degrees), K_o and d^γ are regression coefficients, $K_o = 10$; $d^\gamma = -0.5$, u is a log-normal variable which has zero dB mean and a standard deviation of 8.0 dB.

Using this model, one can observe that the K-factor decreases as the distance increases and as antenna beamwidth increases. It is interesting to determine K-factors that meet the requirements of 90% of all locations within a cell have to be serviced with 99.9% reliability. The calculation of K-factors for this scenario is rather complex, since it also involves path loss, delay spread, antenna correlation (if applicable), specific modem characteristics, and other parameters that influence system performance. However, we can obtain an approximate value as follows: first we select 90% of the users with the highest K-factors over the cell area. Then we obtain the approximate value by selecting the minimum K-factor within the set. For a typical deployment scenario (see the section on SUI channel models) this value of the K-factor can be close to or equal to 0.

2.4.2.4   Doppler Spectrum

Following the Ricean Power Spectral Density (PSD) COST 207, we define scatter and fixed Doppler spectrum components. In fixed wireless channels the Doppler PSD of the scatter (variable) component is mainly distributed around $f = 0$ Hz. The shape of the spectrum is therefore different from the classical Jake spectrum for mobile channels. A rounded shape can be used as a rough approximation to the Doppler PSD, which has the advantage that it is readily available in most existing Radio Frequency (RF) channel simulators. It can be approximated by

$$S(f) = \begin{cases} 1 - 1.72f_0^2 + 0.785f_0^4 & f_0 \leq 1 \\ 0 & f_0 > 1 \end{cases}$$

where $f_0 = \frac{f}{f_m}$. The function is parameterized by a maximum Doppler frequency f_m. Alternatively, the -3 dB point can be used as a parameter, where f_{-3dB} can be related to f_m using the above equation. Measurements at 2.5 GHz center frequency show maximum f_{-3dB} values of about 2 Hz. A better approximation of fixed wireless PSD shapes close to exponential functions. Wind speed combined with foliage (trees), carrier frequency, and the traffic influence the Doppler spectrum. The PSD function of the fixed component is a Dirac impulse at $f = 0$ Hz.

2.4.2.5   Spatial Characteristics, Coherence Distance

Coherence distance is the minimum distance between points in space for which the signals are mostly uncorrelated. This distance is usually greater than 0.5 wavelengths, depending on antenna beamwidth and angle of arrival distribution. At the BTS, it is a common practice to use spacing of about 10 and 20 wavelengths for low-medium and high antenna heights, respectively (120° sector antennae).

2.4.2.6 CCI

The Carrier-to-Interference Ratio (CIR) calculations use a path loss model that accounts for median path loss and log-normal fading, but not for "fast" temporal fading. However, for NLOS cases, temporal fading requires us to allow for a fade margin. The value of this margin depends on the Ricean K-factor of the fading, the *QoS* required, and the use of any fade mitigation measures in the system. Two ways of allowing for the fade margin then arise; either the cumulative distribution function of CIR is shifted left or the CIR required for a non-fading channel is increased by the fade margin. For example, if QPSK requires a CIR of 14 dB without fading, this becomes 24 dB with a fade margin of 10 dB.

2.4.3 Multiple Input Multiple Output (MIMO) Matrix Models

When multiple antennas are used at the transmitter and/or at the receiver, the relationship between transmitter and receiver antennas adds further dimensions to the model. In this case, the channel is characterized not only by the amplitude statistics of each matrix entry (which is usually *Rayleigh* or *Rician*), but also by the *correlation* between these entries (Molisch 2005).

2.4.4 Modified Stanford University Interim (SUI) Channel Models

The channel models described above provide the basis for specifying channels for a given scenario. It is obvious that there are many possible combinations of parameters to obtain such channel descriptions. A set of 6 typical channels is selected for the three terrain types that are typical of the continental United States (Erceg et al. 1999). In this section we present SUI channel models that we have modified to account for 30° directional antennas. These models can be used for simulations, design, development, and testing of technologies suitable for fixed broadband wireless applications. The parameters are selected based upon statistical models described in previous sections. The parametric view of the SUI channels is summarized in Tables 2.3, 2.4, and 2.5. The path loss propagation model in IEEE 802.16 BWA WG (2000) is an experimental model, developed to fit a set of measurements taken in a suburban environment in non-line of sight conditions. As stated in IEEE 802.16 BWA WG (2000), this model is found to fit quite well with models used for urban areas (COST 231-WI) and test drives done in an urban environment. While this model is perfectly adequate for worst case link simulations, it is not adequate for coexistence studies, as it gives quite high estimates for the propagation path loss, and it may underestimate the interference. IEEE 802.16 BWA WG (2000), shows some models where propagation loss is shown as a function of range, predicting about 120 dB path loss for a 1 km range and about a 140 dB

TABLE 2.3
SUI Channel Models for Various Terrains

Terrain Type	SUI Channels
c	SUI-1, SUI-2
b	SUI-3, SUI-4
a	SUI-5, SUI-6

TABLE 2.4
SUI Channels for Low K-Factor

Doppler	Low delay spread	Moderate delay spread	High delay spread
Low	SUI-3	—	SUI-5
High	—	SUI-4	SUI-6

TABLE 2.5
SUI Channels for High K-Factor

Doppler	Low delay spread	Moderate delay spread	High delay spread
Low	SUI-1, 2	—	—
High	—	—	—

loss for $10^{0.8} = 6.3$ km. These values are much larger than the expected path loss in most common cases of rooftop installations, where the propagation conditions are closer to LOS and the receivers are exposed to a much higher interference. There are some alternate models.

2.4.5 FCC Model

The FCC methodology is based upon the basic calculations described in NTIS 2001. The propagation model has three basic elements that affect the predicted field strength at the receiver:

1. LOS mode, using basic free-space path loss

2. NLOS mode, using multiple wedge diffraction

3. Partial first Fresnel zone obstruction losses applicable to either mode

The excess loss component is calculated according to the Epstein–Peterson method.

2.4.6 ITU-R Models

The ITU-R, SG3 has published several recommendations for path loss calculations. This model has the following salient features:

- It takes into account various physical phenomena such as LOS, diffraction, tropospheric scatter, surface ducting, elevated layer reflection and refraction, and hydrometeor scatter.

- It uses the Deygout method, for multiple diffraction.

- Path loss is calculated for clear LOS, LOS with sub-path obstruction and trans-horizon cases.

While the FCC model is focused on the Multichannel Multipoint Distribution Service (MMDS) interference calculation, the ITU-R recommendation is more general in nature and applies for longer range and more diverse cases.

The main drawback from the co-existence study point of view is that the above mentioned models require the ability to calculate the profile between the interferer and the victim, and hence require a digital terrain map of the analysis area. If such a map is not available, or for more general analyses, a simpler model which does not take terrain into account has to be selected. Possible models are

1. Free space propagation

2. Free space models with a variable propagation exponent, clutter constant values, etc.

3. Two-ray, or dual slope models.

2.4.7   Free Space Model

The free space model is the simplest model, but does not model the terrestrial environment reliably. One may heuristically change the coefficient factor, add a constant value according to clutter, etc. However, more theoretical or experimental data are needed to support that.

2.4.8   Two-Ray or Dual Slope Model

This model takes into account the effects of ground reflection and the antenna heights above it. Basically the model takes a free space path loss of 20 dB/decade up to a range $Rb = 4h_{T_x}h_{R_x}/\lambda$, where h_{T_x} and h_{R_x} are the transmitter and receiver antenna heights, respectively, and 40 dB/decade thereafter. This model, although simplistic, can be very well suited for analyses involving LOS scenarios.

2.4.9   Wideband Tapped Delay Line Channel Model

It is mentioned in Hong et al.(2003) that a wideband tapped delay line channel model can be conveniently employed to study a broadband fixed wireless access system at 3.5 GHz. A typical characterization of the channel is shown in Figure 2.4.

2.4.10   Conclusions on Model Selection

The following guidelines may be followed for choosing the appropriate propagation model for co-existence studies:

1. In analyses which include terrain information, the FCC or ITU-R models are recommended.

2. In analyses which do not include terrain information, the FCC or ITU-R models can be used, provided that the model for the terrain profiles can be justified.

3. The "two-ray" model is recommended for simple analyses in which the propagation conditions are clearly LOS.

4. To keep it simple, we can choose one or two models that will be the most conclusive and will cover most common cases.

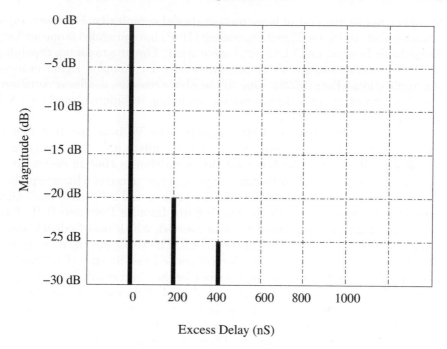

FIGURE 2.4
Wideband Tapped Delay Line Channel Model: Magnitude of Channel Impulse
Response in dB versus Excess Delay in Nanoseconds.

2.5 Classification of Equalizers

Channel equalizers are used in digital communication receivers to mitigate
the effects of Inter-Symbol Interference (ISI) and inter-user interference in
the form of CCI and ACI in the presence of Additive White Gaussian Noise
(AWGN). Linear equalizers based on adaptive filtering techniques have long
been used for this application. Recently, use of nonlinear signal processing
techniques like Artificial Neural Networks (ANN) and Radial Basis Functions
(RBF) have shown encouraging results in this application.

2.5.1 A Note on Historical Development

Early equalizers are based on linear adaptive filter algorithms with or without
a decision feedback. Alternatively, Maximum Likelihood Sequence Estimators
(MLSE) are implemented using the Viterbi algorithm. The linear adaptive
equalizers are simple in structure and easy to train but suffer from poor performance in severe conditions. The infinite memory MLSE provides good per-

formance but at the cost of large computational complexity. Moreover, rapid advancements in Digital Signal Processing (DSP) have provided scope for Very Large Scale Integration (VLSI) implementation. The programming capability of DSP processors makes them very attractive for complex signal processing applications (Patra 1998). Due to the above reasons, nonlinear equalizers have been investigated in the last decade, including techniques based on ANN, RBF and recurrent networks.

The Least Mean Square (LMS) algorithm by Widrow and Hoff (1960) paves the way for the development of adaptive filters used for equalizers. It was Robert W. Lucky who first used this algorithm in 1965 to design adaptive channel equalizers. To overcome the poor performance of a linear equalizer for highly dispersive channels, the MLSE equalizer and its Viterbi implementations were developed in 1970s. The Infinite Impulse Response (IIR) form of the linear adaptive equalizer has also evolved, which uses feedback and is termed a Decision Feedback Equalizer (DFE). The 1980s saw the development of fast convergence algorithms like the Recursive Least Squares (RLS) and the Kalman Filter. Other forms of equalizers like the Fractionally Spaced Equalizer (FSE) were also developed during the period. A review of the development of equalizers up to 1985 can be found in Qureshi (1985).

In the 1990's we saw the development of Multi-Layer Perceptron (MLP) based symbol-by-symbol equalizers. These are computationally more efficient than MLSE and can provide superior performance compared to the conventional equalizers with adaptive filters. The radial basis functions were also used for implementing equalizers subsequently.

The more recent advances in nonlinear equalizers are centered around the application of different signal processing techniques to equalization. Some of these are recurrent neural networks, recurrent RBF, and Mahalonobis classifiers. Currently available equalizers can cater to the needs of CDMA systems as well.

2.5.2 Classification of Adaptive Equalizers

The general equalizer classification is presented in Figure 2.5. In general, the family of adaptive equalizers can be classified as *supervised equalizers* and *unsupervised equalizers*.

The channel distortions introduced into the transmitted signal in the process of transmission can be conveniently removed by transmitting a *training signal* or *pilot signal* periodically during the transmission of information. A replica of this pilot signal is available at the receiver and the receiver uses this to update its parameters during the training period. These kinds of equalizers are known as supervised equalizers. However, the constraints associated with communication systems like Digital Video Broadcast (DVB)/television and Digital Audio Broadcast (DAB) do not provide the scope for the use of a training signal. In this situation the equalizer needs some form of unsupervised or *self-recovery* method to update its parameters so as to provide near

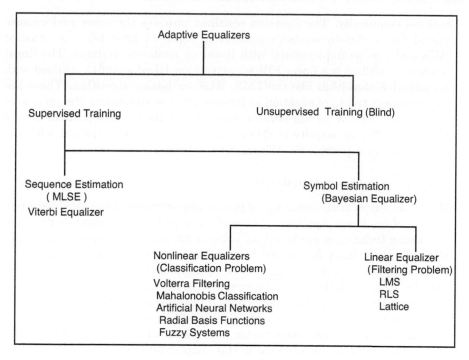

FIGURE 2.5
Classification of Equalizers.

optimal performance. These equalizers are called *blind equalizers*. After training, the equalizer is switched to *decision directed mode*, where the equalizer can update its parameters based on the past detected samples.

The process of supervised equalization can be achieved in two forms. These are *sequence estimation* and *symbol-by-symbol estimation*. Sequence estimator uses the sequence of past received samples to estimate the transmitted symbol. For this reason this form of equalizer is considered as an infinite memory equalizer and is termed MLSE (Forney 1978). The MLSE can be implemented with the Viterbi algorithm (Forney 1973). An infinite memory sequence estimator provides the best Bit Error Rate (BER) performance for equalization of time invariant channels. The symbol-by-symbol equalizer, on the other hand, works as a finite memory equalizer and uses a fixed number of input samples to detect the transmitted symbol. The optimum decision function for this type of equalizer is given by Maximum A Posteriori probability (MAP) criterion and can be derived by Bayes theory. Hence this optimum finite memory equalizer is also called the Bayesian equalizer (Chen et al. 1993). An infinite memory Bayesian equalizer can provide a performance better than the MLSE, but its computational complexity is very large. A finite memory Bayesian equalizer can provide performance comparable to the MLSE with a reduced computational complexity. The Bayesian equalizer provides the lower performance bound for symbol-by-symbol equalizers in terms of probability of error or BER and can be implemented with linear or nonlinear systems. The linear adaptive equalizer is a linear FIR adaptive filter (Haykin 1991) trained with an adaptive algorithm like the LMS, RLS, or lattice algorithm. These linear equalizers treat equalization as inverse filtering and during the process of training optimize a certain performance criteria like Minimum Mean Square Error (MMSE) or amplitude distortion. Linear equalizers trained with the MMSE criterion provide the Wiener filter solution.

2.5.2.1   Nonlinear Equalizers

Recent advances in nonlinear signal processing techniques have provided a rich variety of nonlinear equalizers. Some of the equalizers developed with these processing techniques are based on Volterra filters, ANN, perceptrons, MLP, RBF networks, fuzzy filters, and fuzzy basis functions. All of these nonlinear equalizers, during their training period, optimize some form of a cost function like the MSE or probability of error and have the capability of providing the optimum Bayesian equalizer performance in terms of BER. Nonlinear equalizers treat equalization as a pattern classification process where the equalizer attempts to classify the input vector into a number of transmitted symbols. The fuzzy equalizers investigated in this chapter fall into this category.

Another form of nonlinear equalizer that can be constructed with any of the symbol-by-symbol based equalizers is the DFE, where previously made decisions are used for estimating present and future decisions. This equalizer is also considered an infinite memory equalizer. The conventional DFE using a linear filter is designated a nonlinear equalizer in a wide variety of communication literature, since the decision function used here forms a nonlinear combination of the received samples which is, in fact, the linear combination of the received samples and previously detected samples. In this book, the term nonlinear equalizer is used exclusively for those equalizers that provide a nonlinear decision function based on received samples or received samples along with previously detected samples. The following sections analyze some of the linear and nonlinear equalizers in greater detail.

2.5.3   Optimal Symbol-by-Symbol Equalizer

The optimum symbol-by-symbol equalizer is termed Bayesian equalizer. To derive the equalizer decision function, the discrete time model of the baseband digital communication system presented in Figure 2.6 is considered. The channel is modeled as an FIR filter. The equalizer uses an input vector $r(k) \in \mathcal{R}^m$, the m-dimensional *Euclidean* space. The term m is the equalizer length and the equalizer order can be considered as $(m - 1)$. The equalizer provides a decision function $\mathcal{F}\{r(k)\}$ based on the input vector and this is passed through a decision device to provide the estimate of transmitted signal

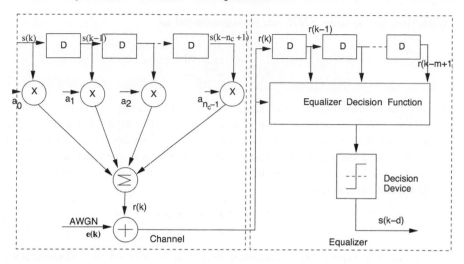

FIGURE 2.6

Block Diagram of a Discrete Time Model of a Digital Communication System (DCS).

$\widehat{s}(k-d)$ where d is the delay associated with equalizer decision. The communication system is assumed to be a two level PAM system where the transmitted sequence $s(k)$ is drawn from an independent identically distributed (i.i.d.) sequence composed of ± 1 symbols. The noise source $\eta(k)$ is assumed to be zero mean, white Gaussian with a variance of σ_η^2. The received signal $r(k)$ at the sampling instant k can be represented as

$$r(k) = \hat{r}(k) + \eta(k) \tag{2.18}$$

$$= \sum_{i=0}^{n_c-1} a_i\, s(k-i) + \eta(k) \tag{2.19}$$

The equalizer performance is described by the probability of misclassification w.r.t. SNR. The SNR is defined as

$$SNR = \frac{E(\hat{r}(k))}{E(\eta(k))} \tag{2.20}$$

$$= \frac{\sigma_s^2 \sum\limits_{i=0}^{n_c-1} a_i^2}{\sigma_n^2} \tag{2.21}$$

where, E is the expectation operator, σ_s^2 represents the signal power, and $\sum\limits_{i=0}^{n_c-1} a_i^2$ is the channel power gain. With the assumption that the signal is drawn from an i.i.d. sequence of ± 1, the variance of signal power $\sigma_s^2 = 1$.

With this, the system SNR can be represented as

$$SNR = \frac{\sum_{i=0}^{n_c-1} a_i^2}{\sigma_n^2} \tag{2.22}$$

The equalizer uses the received signal vector $\mathbf{r}(k) = [r(k), r(k-1), \ldots, r(k-m+1)]^T \in \mathcal{R}^m$ to estimate the delayed transmitted symbol $s(k-d)$. The equalizer with its decision function and a memoryless detector to quantitize the real valued output from decision function $\mathcal{F}\{r(k)\}$ provides an estimate of the transmitted signal. The memoryless detector is implemented using a $sgn(x)$ function given by

$$sgn(x) = \begin{cases} +1, & if \;\; x \geq 0 \\ -1, & if \;\; x < 0 \end{cases} \tag{2.23}$$

The process of equalization discussed here can be viewed as a classification process in which the equalizer partitions the input space into two regions corresponding to each of the transmitted sequences $+1/-1$. The locus of points which separate these two regions is termed the decision boundary. The decision boundary which provides the minimum probability of misclassification is the Bayesian decision boundary derived with the MAP criterion.

2.5.4   Symbol-by-Symbol Linear Equalizers

As discussed before, the linear equalizers in this chapter are equalizers that provide a decision based on the linear combination of the input to the equalizer. If decision feedback is employed, the linear equalizer provides a decision function based on the linear combination of received samples and previously detected samples. The structure of a linear equalizer is presented in Figure 2.7.

The equalizer consists of a TDL which receives the receiver sampled input vector $\mathbf{r}(k) = [r(k), r(k-1), \ldots, r(k-m+1)]^T$ and provides an output $y(k)$ by weighted sum computation of input vector $\mathbf{r}(k)$ with weight vector $\mathbf{w}$. The output is computed once per symbol and can be represented as

$$y(k) = \sum_{i=0}^{m-1} w_i \, r(k-i) \tag{2.24}$$

The weight vector $\mathbf{w}$ optimizes one of the performance criteria like Zero Forcing (ZF) or MMSE criteria. The decision device presented at the output of the filter provides the transmitted signal constellation.

The MMSE criterion provides equalizer tap coefficients $\mathbf{w}(k)$ to minimize the mean square error at the equalizer output before the decision device. This condition can be represented as:

$$J \;\; = \;\; \mathcal{E}|e(k)|^2 \tag{2.25}$$
$$e(k) \;\; = \;\; s(k-d) - y(k) \tag{2.26}$$

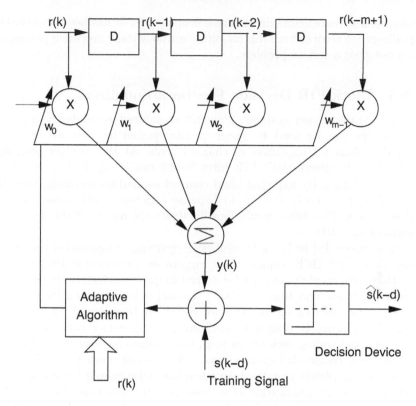

FIGURE 2.7
Block Diagram of a Symbol-by-Symbol Linear Equalizer

where $e(k)$ is the error associated with filter output $y(k)$. The equalizer designed using ZF criterion neglects the effect of noise. However, the MMSE criterion optimizes the equalizer weights for minimizing the MMSE under noise and ISI.The evaluation of the equalizer weights with this criterion requires computation of matrix inversion and the knowledge of the channel, which in most cases is not available. However, adaptive algorithms like LMS (Widrow and Hoff 1960) and RLS can be used to recursively update the equalizer weights during the training period.

A DFE using a linear filter is characterized by its feedforward length m and the feedback order q. The equalizer uses m feed forward samples and q feedback samples from the previously detected samples. The feedback signal vector $\widehat{s}(k) = [\widehat{s}(k-d-1), \widehat{s}(k-d-2), \ldots, \widehat{s}(k-d-q)]^T$ is associated with feedback weight vector $\mathbf{w}_f = [w_0^f, w_1^f, \ldots, w_{q-1}^f]^T$. The feedback section in the equalizer helps to remove the ISI contribution from the estimated symbols. This equalizer provides better performance than the conventional feed forward linear equalizer. When there is an error in the decision, the error is fed back and this results in more errors due to error propagation. It has been observed that equalizers can recover from this condition automatically and error propagation does not pose a serious problem.

2.5.5 Block FIR Decision Feedback Equalizers

In block transmission systems, transmitter-induced redundancy using FIR filter banks can be used to suppress intersymbol interference and equalize FIR channels irrespective of channel zeros. At the receiver end, linear or Decision Feedback (DF) FIR filter banks can be applied to recover the transmitted data. By applying blind channel estimation methods, filter bank transmitter-receivers (transceivers) dispense with bandwidth consuming training sequences. Extensive simulations illustrate the merits of the design (Stamoulis et al. 2001).

To suppress ISI in block transmission systems, transmission precoding is done along with DFE. Equalization targets such structured ISI-induced errors that are caused by multipath-induced frequency-selective channels. If the (presumed linear and time-invariant) channel is known, then its structured, deterministic effect on the transmitted signal can be removed (or significantly reduced) by properly designed equalizers at the receiver end. On the other hand, channel coding techniques (e.g., convolutional codes) are effective for unstructured (noise-like) symbol errors. As a result, even when the channel cannot be completely equalized, or when the noise cannot be suppressed (as with zero-forcing equalization of a channel with nulls close to the unit circle), channel coding lowers (but does not remove) the error floor in the BER performance at the expense of introducing redundancy.

To combat fading effects in frequency selective channels, the transmitter does not have channel coding only at its disposal. Redundant block transmission systems such as Orthogonal Frequency Division Multiplexing (OFDM) rely on Inverse Fast Fourier Transform (IFFT) precoding to cope with ISI. Among the ways to model block transmission of data is the unifying framework that enables most of the currently used block transmission systems to be realized using pairs of filter bank transmitters and receivers. The introduction of very modest redundancy relative to channel coding and transmitter precoding also enables blind channel estimation and block synchronization. The redundancy is in the form of cyclic prefix or zero padding (which acts as a guard interval) and offers degrees of freedom that can be exploited when designing transceivers under BER and information rate (throughput) constraints.

However, the BER performance of the equalization process depends critically upon the receiver structure. Serial DF receivers have been shown to exhibit superior BER performance (when compared to linear receivers) and have the potential to achieve (under certain conditions) the performance of the maximum-likelihood receiver. Moreover, with adaptive DFE techniques, the DFE receiver structure lends itself naturally to decision-directed channel estimation. Blind DFE channel estimation methods have also been proposed.

As their name suggests, serial DF receivers apply the same filters to every received symbol. Though serial DF receivers can be used in block transmission systems, they do not fully exploit the structure of the received blocks. On the other hand, block DF receivers apply different filters to symbols of the received block and can result in improved BER performance. Unlike serial ZF DF receivers, which entail IIR feedforward and feedback structures, it is shown that block ZFDF receivers are given by closed-form expressions, which can be implemented exactly using FIR filter banks (Stamoulis et al. 2001).

The block FIR DF receivers can be realized exactly and outperform the hybrid block/serial DF receiver structures proposed recently for OFDM transmissions (Stamoulis et al. 2001). Note that in the DF framework, the decision device produces only one symbol estimate at a time. This in contrast to the block DF framework, where the decision device can be tuned to collect a block of received symbols, produces symbol estimates for the corresponding transmitted symbols, and asymptotically achieves the performance of the maximum-likelihood receiver.

2.5.6 Symbol-by-Symbol Adaptive Nonlinear Equalizer

Nonlinear equalizers treat equalization as a nonlinear pattern classification problem and provide a decision function that partitions the input space $\mathcal{R}^n$ to the number of partitions each associated with one transmitted symbol. As a result the equalizer assigns the input vector to one of the signals in the constellation. The nonlinear equalizers introduced in this section are based on the RBF networks and the ANN. Some of the other forms of nonlinear equalizers are those based on the recurrent RBF (Cid-Sueiro et al. 1994), the recurrent ANN (Parishi et al. 1997), the Volterra filters, the functional link network, and Mahalonobis classifiers.

2.5.6.1 RBF Equalizer

The RBF network was originally developed for interpolation in multidimensional space. A schematic of this RBF network with m inputs and a scalar output is presented in Figure 2.8. This network can implement a mapping $f_{rbf} : \mathcal{F} \to \mathcal{R}^m$ by the function

$$f_{rbf}\{\mathbf{x}(k)\} = \sum_{i=1}^{N_r} w_i \phi(\|\mathbf{x}(k) - \rho_i\|) \qquad (2.27)$$

where $\mathbf{x}(k) \in \mathcal{R}^m$ is the input vector, $\phi(.)$ is the given function from $\mathcal{R}^+$ to $\mathcal{R}$, w_i, $1 \leq i \leq N_r$ are weights, and $\rho_i \in \mathcal{R}^m$ are known as RBF centers. This RBF structure can be extended for multidimensional output as well. Possible choices for the radial basis function include a thin plate spline, $\phi(\gamma) = \frac{\gamma}{\sigma_r^2} \log\left(\frac{\gamma}{\sigma_r}\right)$, a multiquadratic, $\phi(\gamma) = \sqrt{\gamma^2 + \sigma_r^2}$, an inverse multiquadratic, $\phi(\gamma) = \frac{1}{\sqrt{\gamma^2 + \sigma_r^2}}$, and the Gaussian kernel, $\phi(\gamma) = \exp\left(-\frac{\gamma^2}{2\sigma_r^2}\right)$.

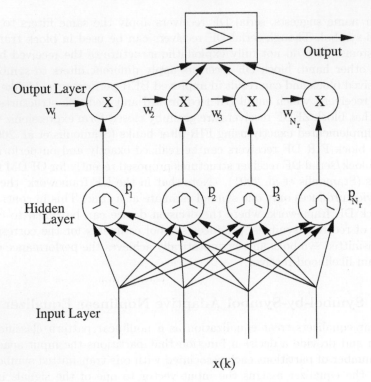

FIGURE 2.8
Block Diagram of a Radial Basis Function Neural Network (RBF NN).

Here, the parameter σ_r^2 controls the radius of influence of each basis function and determines how rapidly the function approaches 0 with γ (Mulgrew 1996).The Gaussian and the inverse multiquadratic kernel provide bounded and localized properties such that $\phi(\gamma) \to 0$ as $\gamma \to \infty$. Broomhead and Lowe reinterpret the RBF network as a least square estimator, which led to its widespread use in signal processing applications such as time series prediction, system identification, interference cancellation, radar signal processing, pattern classification, and channel equalization. In signal processing applications the RBF inputs are presented through a TDL. Training of the RBF networks involves setting the parameters for the centers ρ_i, spread σ_r and the linear weights w_i. The RBF networks are easy to train since the training of centers, spread factor, and the weights can be done sequentially and the network offers a nonlinear mapping, maintaining its linearity in parameter structure at the output layer. One of the most popular schemes employed for training the RBF in a supervised manner is to estimate the centers using a clustering algorithm like κ-means clustering and setting σ_r^2 to an estimate of input noise variance calculated from the center estimation error. The output layer weights can be trained using the popular stochastic gradient LMS algorithm. Other schemes for RBF training involve selecting a large number of candidate centers initially and using the orthogonal least squares (OLS) algorithm to pick a subset of the centers that provides near optimal performance. The MLP back propagation algorithm can also be used to train the RBF centers (Mulgrew 1996).

In early RBF equalizers, the RBF centers are selected at random, picked from a few of the initial input vectors. The weights are updated using supervised training by the LMS algorithm or its momentum version. This results in equalizers with a large number of centers, making the network computationally complex. Chen proposes the OLS algorithm for selecting an optimum number of centers from a large number of candidate centers, resulting in near optimal performance (Mulgrew 1996). Subsequently, the close relationship between the RBF network and the Bayesian equalizer is found and this provides the parametric implementation of the Bayesian equalizers with the RBF. In these equalizers, supervised κ-means clustering provides the estimate of the centers while linear weights are estimated using the LMS algorithm. With the development of RBFs that could handle complex signals, they are used for equalization in communication systems with complex signal constellation. Cha proposed the stochastic gradient algorithm to adapt all the RBF parameters and used this technique to equalize 4-QAM digital communication systems (Mulgrew 1996).

A deeper examination of the RBF decision function in Equation 2.27, in conjunction with a Gaussian kernel, and the Bayesian equalizer decision function shows that both of these functions are similar. The RBF network can provide a Bayesian decision function by setting the RBF centers, ρ_i, to channel states, $\mathbf{c}_i$, RBF spread parameter, σ_r^2, to channel noise variance, σ_η^2, and the linear weights $w_i = +1 \ if \ \mathbf{c}_i \in \mathbf{C}_d^+$ and $w_i = -1 \ if \ \mathbf{c}_i \in \mathbf{C}_d^-$. This provides the optimum RBF network as an equalizer. In this implementation the channel state vectors $\mathbf{c}_i$ can be estimated using supervised κ-means clustering or alternatively they can be calculated from an estimate of the channel.

RBF equalizers can provide optimal performance with small training sequences but they suffer from computational complexity. The number of RBF centers required in the equalizer increases exponentially with equalizer order and the channel delay dispersion order. In a varied implementation, the RBF with scalar centers results in a reduction of computational complexity. The issues relating to RBF equalizer design have been discussed extensively in (Mulgrew 1996).

2.5.6.2 Fuzzy Adaptive Equalizer (FAE)

The fuzzy adaptive filter is constructed from a set of fuzzy IF-THEN rules that change adaptively to minimize some criterion function as new information becomes available. The concept can be generalized to include complex parameters and complex signals (Lee 1996). The fuzzy filter as adaptive equalizer is applied to Quadrature Amplitude Modulation (QAM) digital communication with linear complex channel characteristics. The fuzzy adaptive filter has drawn a great deal of attention because of its universal approximation ability in nonlinear problems. These fuzzy rules come either from human experts or by matching input-output pairs through an adaptation procedure (Lee 1996). Some application examples of the fuzzy filter to signal processing include classification and signal prediction, communications channel equalization, and nonlinear systems modeling and identification. Most fuzzy filters available to us are real-valued and are suitable for signal processing in real multidimensional space. In some applications, however, signals are complex valued and processing is done in complex multidimensional space. An example is the equalization of digital communication channels with complex signaling schemes such as QAM. For complex signal processing problems, many existing fuzzy filters cannot be directly applied.

The complex fuzzy adaptive filter with changeable fuzzy IF-THEN rules is an extension of the real fuzzy filter. The inputs and outputs as well as the parameters of the filter are all complex-valued. However, the membership function of this is real. The filter can be viewed as a mapping from the complex multi-input onto the complex single-output. To adjust coefficients and the parameters of the membership functions that characterize the fuzzy concepts in the IF-THEN rules, the adaptive algorithm based on LMS is used. When both the filter inputs and desired outputs are reduced to real-valued, this complex fuzzy filter degenerates naturally into the real fuzzy filter. Also, to demonstrate the effectiveness of this algorithm, the complex fuzzy adaptive filter as equalizer is applied to a linear channel equalization problem based on a four-QAM scheme.

2.5.6.3　Equalizer Based on Feedforward Neural Networks

A signal suffers from nonlinear, linear, and additive distortion when transmitted through a channel. Linear equalizers are commonly used in receivers to compensate for linear channel distortion. As an alternative, nonlinear equalizers have the potential to compensate for all three sources of channel distortion (Lu and Evans 1999). It has been shown that nonlinear feedforward equalizers based on either Multi-Layer Perceptron (MLP) or RBF neural networks can outperform linear equalizers. A reduced complexity neural network equalizer can be built by cascading an MLP and an RBF network. In simulation, the new MLP-RBF equalizer outperformed MLP equalizers and RBF equalizers in symbol error rate versus SNR (Lu and Evans 1999).

Equalization may either require a training signal or be blind. In digital communications, the training signal is simply a known sequence of symbols sent by transmitter so that the receiver can estimate the channel distortion. Linear equalizers that employ training sequences are often based on adaptive FIR filters. They are easy to implement and track linear distortion in the channel fairly well, provided that enough taps are used (using 50 to 100 taps is common). Some linear equalizers, such as a zero-forcing equalizer, may amplify channel noise. As an alternative, nonlinear equalizers have the potential to compensate for all three sources of channel distortion. A common nonlinear equalizer is the DFE. Another class of nonlinear equalizers is based on artificial neural networks, e.g., MLP and RBF feedforward neural networks.

Comparison of the symbol error rate versus SNR performance of MLP, RBF, and MLP-RBF equalizers using different channel characteristics, number of input neurons, and the number of hidden neurons shows that the new structure is

1. A reduced complexity MLP-RBF neural network equalizer.

2. Its performance is better than that of MLP equalizers and RBF equalizers.

2.5.6.4   A Type-2 Neuro Fuzzy Adaptive Filter

This is a *model free approach*. Type-2 fuzzy sets have grades of membership that are themselves fuzzy. A type-2 membership grade can be any subset in [0,1], the primary membership. Type-2 fuzzy sets allow us to handle linguistic uncertainties. The implementation of the adaptive filter is discussed in references Savazzi et al. (1998), Patil and Singh (2004), Liang and Mendel (2000), Wang and Mendel (1992, 1993). The type-2 FAF is discussed in detail in Chapter 3.

2.5.7   Equalizer Based on the Nearest Neighbor Rule

Performance degradation in a mobile radio communication system is due to physical phenomena, such as multipath fading and time and Doppler delay spread that produce ISI and to the variations in the time of the CIR. To counteract the impairments in CIR, the Global System for Mobile communications (GSM) uses an MLSE receiver, based on the Viterbi algorithm. This algorithm is well known to be the optimum solution for detecting an information sequence corrupted by ISI and additive Gaussian noise. Its optimality is based on the assumption that the statistical behavior of the channel is known. When this assumption does not hold, as in mobile communication applications where the environment changes not only for each different connection but also within the same call, channel estimation is needed to derive the correct metrics to be used in the evaluation of the decoding path. Adaptive channel estimation is performed by inserting known training sequences into the transmitted information and then mapping the alterations of the known bit patterns into new metrics for the information part of the transmitted burst. The complexity of this approach is known to grow exponentially with the system memory, as this determines the number of states to be used in the trellis diagram. This has led to the exploitation of pruning algorithms to reduce either the state space or the number of paths tracked along the trellis, trading part of the performance for a reduction in complexity (Savazzi et al. 1998).

Recent applications of clustering and neural network techniques to channel equalization have revealed the classification nature of this problem. The most important advantage in using the nearest neighbor classification algorithm is the significant reduction in terms of computational complexity compared with the MLSE equalizer.

The proposed approach involves symbol-by-symbol interpretation and the knowledge of the channel is embedded in the mapping process of the received symbols over the symbols of the training sequence. This means that no explicit channel estimation need be carried out, either with correlative blocks or using neural networks, thus speeding up the entire process. The performance of the proposed receiver, evaluated through a channel simulator for mobile radio communications, is compared with the results obtained by means of a 16-state Viterbi algorithm and other suboptimal receivers. It is shown that the NN classification algorithm increases the BER compared with the MLSE demodulator, but the performance degradation, despite the simplicity of the receiver, is kept within the limits imposed by the GSM specifications.

2.6 Conclusion

In this chapter, the basic concepts in mobile cellular communications are introduced. It is shown how the *frequency reuse* will result in *CCI* and how it can be kept low by increasing the cluster size, N. The need for a channel equalizer in combating CCI is introduced. Then different channel models used in simulating the mobile channel are considered. This is followed by a detailed study of various channel equalizers developed to date. The working principles of a number of channel equalizers are also considered.

Further Reading

Andreas F. Molisch, *Wireless Communications*, Sponsored by IEEE Press, John Wiley, New York 2005.

Anastasios Stamoulis, Georgios B. Giannakis, and Anna Scaglione, Block FIR Decision-Feedback Equalizers for Filterbank Precoded Transmissions with Blind Channel Estimation Capabilities, *IEEE Transactions on Communications*, Vol. 49, No. 1, January 2001.

B.P. Patil and J. Singh, Adaptive DQPSK under Fading Statistics for Mobile Communication, *Journal of the Institution of Engineers (India)*, Vol.84, January 2004.

B. Widrow and M.E. Hoff, Jr., Adaptive Switching Circuits, *Proceedings of IRE WESCON Convention*, Vol.4, pp.94–104, August 1960.

B. Mulgrew, Applying Radial Basis Functions, *IEEE Signal Processing Magazine*, Vol.13, pp.50–65, March 1996.

Biao Lu and Brian L. Evans, Channel Equalization by Feedforward Neural Networks, *Proceedings of IEEE International Symposium on Circuits and Systems*, May 30-Jun. 2, 1999, Orlando, FL, Vol.5, pp. 587–590.

Chia Leong Hong et al., Wideband Tapped Delay Line Channel Model at 3.5GHz for Broadband Fixed Wireless Access System as Function of Subscriber Antenna Height in Suburban Environment, *Proceedings of ICICS-PCM 2003*, 1B7.4, Singapore, December 2003.

D.S. Broomhead and D. Lowe, Multivariate Functional Interpolation and Adaptive Networks, *Complex Systems*, Vol. 2, pp. 321–355, 1988.

G.D. Forney, The Viterbi Algorithm, *Proceedings of the IEEE*, Vol.61, pp.268–278, March 1973.

G.D. Forney, Maximum-Likehood Sequence Estimation of Digital Sequence in the Presence of Inter-Symbol Interface, *IEEE Transactions on Information Theory*, Vol. IT-18, pp. 363–378, May 1978.

IEEE Working Group, Channel Models for Broadband Fixed Wireless Systems, *IEEE 802.16 Broadband Wireless Access Working Group Document*, 2000.

J. Cid-Sueiro, A. Artes-Roddriguez, and A.R. Figueiras-Vidal, Recurrent Radial Basis Function Network for Optimal Symbol-by-Symbol Equalisation, *Signal Processing (Eurasip)*, Vol.40, pp.53–63, October 1994.

John G. Proakis and Masoud Salehi, *Communication Systems Engineering*, 2nd edition, Pearson Education, New Jersey 2002.

Ki Yong Lee, Complex Fuzzy Adaptive Filter with LMS Algorithm, *IEEE Transactions on Signal Processing*, Vol.44, No.2, pp.424–427, February 1996.

Li-Xin Wang and Jerry M. Mendel, Fuzzy Basis Functions, Universal Approximation, and Orthogonal Least-Squares Learning, *IEEE Transactions on Neural Networks*, Vol.3, No.5, pp.807–814, September 1992.

Li-Xin Wang and Jerry M. Mendel, An RLS Fuzzy Adaptive Filter, with Application to Nonlinear Channel Equalization, *IEEE Transactions on Fuzzy Systems*, Vol.1, pp.895–900, August 1993.

L.J. Greenstein, V. Erceg, Y.S. Yeh, and M.V. Clark, A New Path-Gain/Delay-Spread Propagation Model for Digital Cellular Channels, *IEEE Transactions on Vehicular Technology*, Vol. 46, No. 2, May 1997.

L.J. Greenstein, S. Ghassemzadeh, V. Erceg, and D.G. Michelson, Ricean K-factors in Narrowband Fixed Wireless Channels: Theory, Experiments, and Statistical Models, *Proceedings of WPMC Conference*, Amsterdam, September 1999.

M. Hata, Empirical Formula for Propagation Loss in Land Mobile Radio Services, *IEEE Transactions on Vehicular Technology*, Vol. 29, pp. 317–325, August 1980.

M.S. Smith and J.E.J. Dalley, A New Methodology for Deriving Pathloss Models from Cellular Drive Test Data, *Proceedings of AP2000 Conference*, Davos, Switzerland, April 2000.

NTIS, Transmission Loss Prediction for Tropospheric Communication Circuits, *Technical Note 101, NTIS Access Number AD 687-820*, National Technical Information Service, US Department of Commerce, Springfield, VA 2001.

Pietro Savazzi, Lorenzo Favalli, Eugenio Costamagna, and Alessandro Mecocci, A Suboptimal Approach to Channel Equalization Based on the Nearest Neighbor Rule, *IEEE Journal of Selected Areas in Communications*, Vol.16, No.9, pp.1640–1648, December 1998.

Qilian Liang and Jerry M. Mendel, Equalization of Nonlinear Time-Varying Channels Using Type2 Fuzzy Adaptive Filters, *IEEE Transactions on Fuzzy Systems*, Vol.8, No.5, pp.551–563, October 2000.

Qilian Liang and Jerry M. Mendel, Overcoming Time-Varying Co-Channel Interference Using Type2 Fuzzy Adaptive Filters, *IEEE Transactions on Circuits and Systems II:Analog and Digital Signal Processing*, Vol.47, No.12, pp.1419–1428, December 2000.

R. Parishi, E.D.D. Claudio, G. Orlandi, and B.D. Rao, Fast Adaptive Digital Equalization by Recurrent Neural Networks, *IEEE Transactions on Signal Processing*, Vol.45, pp.2731–2739, November 1997.

Roger L. Peterson, Rodger E. Ziemer, and David E. Borth, *Introduction to Spread Spectrum Communications*, Pearson Education, New Jersey 1995.

Sarat Kumar Patra, Development of Fuzzy System Based Channel Equalisers, Ph.D. thesis, University of Edinburgh, August 1998.

S.U.H. Qureshi, Adaptive Equalization, *Proceedings of the IEEE*, Vol.73, pp.1349–1387, September 1985.

S. Chen, B. Mulgrew,and S. McLaughlin, Adaptive Bayesian Equalizer with Decision Feedback, *IEEE Transactions on Signal Processing*, Vol.41, pp.2918–2927, September 1993.

Simon Haykin, *Adaptive Filter Theory*,Prentice Hall, New Jersey 1991.

Theodore S. Rappaport, *Wireless Communications Principles and Practice*, Pearson Education, New Jersey 2003.

Theodore S. Rappaport and Lawrence B. Milstein, Effects of Radio Propagation Path Loss on DS-CDMA Cellular Frequency Reuse Efficiency for the Reverse Channel, *IEEE Transactions on Vehicular Technology*, Vol.41, No.3, pp.231–242, August 1992.

V. Erceg et al., An Empirically Based Pathloss Model for Wireless Channels in Suburban Environments, *IEEE Journal of Selected Areas in Communication*, Vol. 17, No. 7, July 1999, pp.1205–1211.

William C.Y. Lee, Elements of Cellular Mobile Radio Systems, *IEEE Transactions on Vehicular Technology*, Vol.VT-35, No.2, pp.48–56, May 1986.

Y. Okumura, E. Ohmori, T. Kawano, and K. Fukua, Field Strength and Its Variability in UHF and VHF Land-Mobile Radio Service, *Review Electrical Communication Lab.*, Vol. 16, No. 9, 1968.

3

Neuro-Fuzzy Equalizers for Cellular Channels

In the previous chapter, we have seen that the mobile cellular channel is *Linear and Time Variant (LTV)*, because of fading (Liang and Mendel 2000). Therefore an equalizer which can be effectively applied to such a channel needs to be nonlinear, as there are certain uncertainties in the channel parameters (Adali 1999). Since the mobile cellular channel is *broadcast type*, blind equalizers are most suitable for them (Xie and Leung 2005). We consider mainly two equalizers in this chapter. They are the *Type-2 Fuzzy Adaptive Filter* based Transversal Equalizer (TE) and the Decision Feedback Equalizer (DFE). Decision feedback has been used both in linear and nonlinear equalizers to improve system performance (Liang and Mendel 2000).

The rest of the chapter is organized as follows. In Section 3.1, we start with the fundamental concepts of Neuro-Fuzzy systems. The Fuzzy Adaptive Filter (FAF) is introduced in Section 3.2, followed by an adaptation of the same for equalizers for indoor mobile cellular channels in Section 3.3. In Section 3.4, we compare it with conventional designs and arrive at important conclusions.

3.1 Introduction to Neuro-Fuzzy Systems

Fuzzy logic and neural networks (with genetic algorithms) are complementary technologies in the design of intelligent systems (Lin and Lee 1996). Fuzzy logic is based on the way the brain deals with inexact information, while neural networks are modeled after the physical architecture of the brain. Fuzzy systems and neural networks are both numerical *model-free estimators* and dynamic systems. They share the common ability to improve the intelligence of systems working in an uncertain, imprecise, and noisy environment. Neural networks provide fuzzy systems with learning abilities, and fuzzy systems provide neural networks with a structured framework with high-level fuzzy *IF–THEN* rule thinking and reasoning.

3.1.1   Fuzzy Systems and Type-1 Fuzzy Sets

A classical (crisp) set is a collection of distinct objects. A crisp set can be defined by the *characteristic function*. Let U be the universe of discourse. The *characteristic function* $\mu_A(x)$ of a crisp set A in U takes its values in $\{0,1\}$ and is defined as

$$\mu_A(x) = \begin{cases} 1 & if \quad and \quad only \quad if \quad x \in A \\ 0 & if \quad and \quad only \quad if \quad x \notin A. \end{cases} \tag{3.1}$$

A type-1 fuzzy set, on the other hand, introduces vagueness by eliminating the sharp boundary that divides members from nonmembers in the group. A fuzzy set $\tilde{A}$ in the universe of discourse U can be defined as a set of ordered pairs,

$$\tilde{A} = \{(x, \mu_{\tilde{A}}(x)) | x \in U\}, \tag{3.2}$$

where $\mu_{\tilde{A}}(.)$ is called the *membership function* (or characteristic function) of $\tilde{A}$ and $\mu_{\tilde{A}}(x)$ is the *grade* (or degree) of membership of x in $\tilde{A}$, which indicates the degree that x belongs to $\tilde{A}$ (Lin and Lee 1996). Note that $\mu_{\tilde{A}}(x)$ can take any value in the closed interval $[0,1]$.

3.1.2   Type-2 Fuzzy Sets

A fuzzy set whose membership function is itself a fuzzy set is called a *type-2 fuzzy set*. A *type-1 fuzzy set* is an ordinary fuzzy set. Hence a type-2 fuzzy set is a fuzzy set whose membership values are type-1 fuzzy sets on $[0,1]$ (Lin and Lee 1996, Zadeh 1965). A type-2 fuzzy set in a universe of discourse U is characterized by a *fuzzy membership function* μ_A as

$$\mu_A : U \longrightarrow [0,1]^{[0,1]}, \tag{3.3}$$

where $\mu_A(x)$ is the *fuzzy grade* and is a fuzzy set in $[0,1]$ represented by

$$\mu_A(x) = \int f(u)/u, \quad u \in [0,1], \tag{3.4}$$

where f is a membership function for the fuzzy grade $\mu_A(x)$ and is defined as

$$f : [0,1] \longrightarrow [0,1]. \tag{3.5}$$

Type-2 Membership Functions are illustrated in Figure 3.1.

3.1.2.1   Extension Principle

The *extension principle*, introduced by Zadeh, is one of the most important tools of fuzzy set theory. Using the extension principle, any mathematical relationship between nonfuzzy elements can be extended to deal with fuzzy entities (Lin and George Lee 1996). The extension principle is stated as follows.

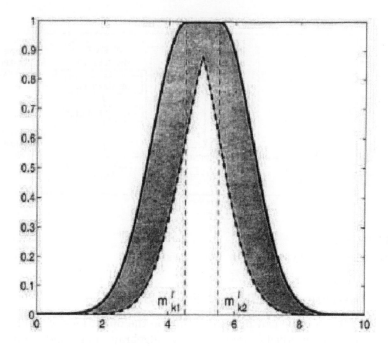

FIGURE 3.1
Type-2 Membership Functions.

Given a function $f : U \longrightarrow V$ and a fuzzy set A in U, where $A = \mu_1/x_1 + \mu_2/x_2 + \ldots + \mu_n/x_n$,[1] the extension principle states that

$$
\begin{aligned}
f(A) &= f(\mu_1/x_1 + \mu_2/x_2 + \ldots + \mu_n/x_n) \\
&= \mu_1/f(x_1) + \mu_2/f(x_2) + \ldots + \mu_n/f(x_n)
\end{aligned} \tag{3.6}
$$

If more than one element of U is mapped to the same element y in V by f (i.e., a many to one mapping), then the maximum among their membership grades is taken. That is,

$$
\mu_{f(A)}(y) = \max_{\substack{x_i \in U \\ f(x_i)}} [\mu_A(x_i)], \tag{3.7}
$$

where x_i are the elements that are mapped to the same y (Lin and Lee 1996).
The structure of a type-2 Fuzzy Logic System is shown in Figure 3.2.

---

[1]The + signs indicate logical *OR*–ing.

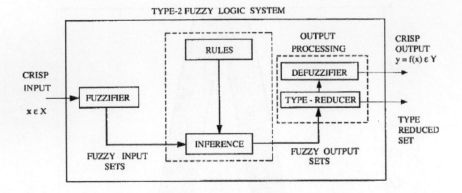

FIGURE 3.2
The Structure of a Type-2 Fuzzy Logic System.

3.1.3   Operations on Type-2 Fuzzy Sets

The extension principle can be used to define the operations of intersections
(e.g., an algebraic product), union (e.g., an algebraic sum), and complement
of type-2 fuzzy sets. Let $\mu_A(x)$ and $\mu_B(x)$ be fuzzy grades for type-2 fuzzy
sets A and B, respectively, and they are defined by

$$\mu_A(x) = \int f(u)/u, \ \ u \in [0,1], \quad \mu_B(x) = \int g(w)/w, \ \ w \in [0,1], \quad (3.8)$$

where f and g depend on x as well as on u or w. Using the extension principle,
we have the following:

1. Min operator $[A \cap B]$:

$$\mu_{A \cap B}(x) \ = \ \mu_A(x) \cap \mu_B(x) = \int f(u)/u \cap \int g(w)/w$$

$$= \ \int f(u) \wedge g(w)/u \wedge w; \ \ where \ \wedge \ denotes \ min. \ (3.9)$$

2. Max operator $[A \cup B]$:

$$\mu_{A \cup B}(x) \ = \ \mu_A(x) \cup \mu_B(x) = \int f(u)/u \cup \int g(w)/w$$

$$= \ \int f(u) \wedge g(w)/u \vee w; \ \ where \ \vee \ denotes \ max. \ (3.10)$$

3. Algebraic product $[AB]$:

$$\mu_{AB}(x) = \mu_A(x).\mu_B(x) = \int f(u)/u \int g(w)/w$$

$$= \int f(u) \wedge g(w)/uw. \tag{3.11}$$

4. Algebraic sum $[A \overset{\wedge}{+} B]$:

$$\mu_{A\overset{\wedge}{+}B}(x) = \mu_A(x) \overset{\wedge}{+} \mu_B(x) = \int f(u) \wedge g(w)/u \overset{\wedge}{+} w$$

$$= \int f(u) \wedge g(w)/u + w - uw. \tag{3.12}$$

5. Complement $[\overline{A}]$:

$$\mu_{\overline{A}}(x) = \overline{\mu_A(x)} = \int f(u)/(1-u). \tag{3.13}$$

The operations on type-2 fuzzy sets are discussed in Karnik et al. (1999) and Mendel (2000).

3.2 Type-2 Fuzzy Adaptive Filter

The type-2 fuzzy adaptive filter for channel equalization is obtained by generalizing the unnormalized output type-1 Takagi–Sugeno–Kang (TSK) fuzzy logic system to a type-2 TSK fuzzy logic system (Liang and Mendel 2000) For equalization, the antecedents of type-1 TSK FLS are generalized to type-2 fuzzy sets, whereas the consequent is unchanged (i.e., it is a crisp number).

In a type-2 FAF with a rule base of M rules, where each rule has p antecedents, the ith rule R^i is denoted as

$$R^i : IF \ x_1 \ is \ \tilde{F}_1^i \ and \ x_2 \ is \ \tilde{F}_2^i \ and \ldots and \ x_p \ \tilde{F}_p^i \tag{3.14}$$

$$THEN \ y^i = c_0^i + c_1^i x_1 + c_2^i x_2 + \ldots + c_p^i x_p$$

where $i = 1, 2, \ldots, M; c_j^i (j = 0, 1, \ldots, p)$ are the consequent parameters that are crisp numbers; y^i is an output from the IF–THEN rule, which is a crisp number; and the $\tilde{F}_k^i (k = 1, 2, \ldots, p)$ are type-2 fuzzy sets. Given an input $\mathbf{x} = [x_1, x_2, \ldots, x_p]^T$, the firing strength of the ith rule is

$$F^i = \mu_{\tilde{F}_1^i}(x_1) \sqcap \mu_{\tilde{F}_2^i}(x_2) \sqcap \ldots \sqcap \mu_{\tilde{F}_p^i}(x_p) \ where \ \sqcap \ denotes \ t-norm. \tag{3.15}$$

The final output of the type-2 FAF is obtained by applying the *extension principle* to Equation 3.16.

$$y = \sum_{i=1}^{M} f^i y^i \tag{3.16}$$

Thus the final output is given by

$$Y(F^1, \ldots, F^M) = \int_{f^1} \cdots \int_{f^M} \mathcal{T}_{i=1}^M \mu_{F^i}(f^i) \Bigg/ \sum_{i=1}^{M} f^i y^i \tag{3.17}$$

where M is the number of rules fired, $f^i \in F^i$, and $\mathcal{T}$ indicates the chosen t-norm. Y is called an *extended weighted average*; it reveals the uncertainty at the output of a type-2 FLS due to antecedent uncertainties and is itself a type-1 fuzzy set.

When interval type-2 sets are used in the antecedents, which means $\mu_{\tilde{F}_k^i}(x_k) \ (k = 1, \ldots, p)$ is an interval set, we write

$$\mu_{\tilde{F}_k^i}(x_k) = \left[\underline{\mu}_{\tilde{F}_k^i}(x_k), \overline{\mu}_{\tilde{F}_k^i}(x_k)\right] \triangleq \left[\underline{f}_k^i, \overline{f}_k^i\right]. \tag{3.18}$$

The type-2 FAF is computed using results in the following steps:

1. In an interval type-2 FAF, which will meet under minimum or product t-norm, the firing strength in Equation 3.15 for rule R^i is an interval set $F^i = [\underline{f}^i, \overline{f}^i]$, where $(i = 1, \ldots, M)$

$$\underline{f}^i = \underline{\mu}_{\tilde{F}_1^i}(x_1) \star \ldots \star \underline{\mu}_{\tilde{F}_p^i}(x_p) = \mathcal{T}_{k=1}^p \underline{f}_k^i \tag{3.19}$$

and

$$\overline{f}^i = \overline{\mu}_{\tilde{F}_1^i}(x_1) \star \ldots \star \overline{\mu}_{\tilde{F}_p^i}(x_p) = \mathcal{T}_{k=1}^p \overline{f}_k^i. \tag{3.20}$$

2. The extended weighted average $Y(F^1, \ldots, F^M)$ is also an interval set $[y_l, y_r]$ where

$$y_r = \sum_{1=1}^{M} \overline{f}^i y^i \tag{3.21}$$

$$y_l = \sum_{1=1}^{M} \underline{f}^i y^i \tag{3.22}$$

and

$$y^i = c_0^i + c_1^i x_1 + c_2^i x_2 + \ldots + c_p^i x_p. \tag{3.23}$$

3. The defuzzified output of the type-2 FAF is

$$y = \sum_{i=1}^{M} y^i (\underline{f}^i + \overline{f}^i)/2. \tag{3.24}$$

3.2.1 TE for Time-Varying Channels

The structure of a TE with p taps is given in Figure 3.3. If $s(k)$ is the symbol

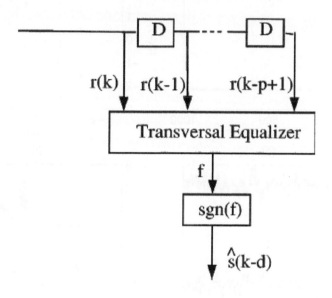

FIGURE 3.3
The Structure of the TE.

transmitted, $e(k)$ is the noise, the channel order is n ($n + 1$ *taps*), and the time-varying tap coefficients are $a_i(k)$ ($i = 0, 1, \ldots, n$) then the received symbol, $r(k)$, can be represented as

$$r(k) = \sum_{i=0}^{n} a_i(k)s(k - i) + e(k). \tag{3.25}$$

We assume that $s(k)$ is binary, i.e., it is either $+1$ or -1 with equal probability.
If p is the TE order (i.e., number of taps in the equalizer), we denote

$$\mathbf{r(k)} \triangleq [r(k), r(k - 1), \ldots, r(k - p + 1)]^T. \tag{3.26}$$

Note that $\mathbf{r(k)}$ depends on the channel input sequence $\mathbf{s(k)}$ (an $(n + p) \times 1$ vector), where

$$\mathbf{s(k)} = [s(k), s(k - 1), \ldots, s(k - n - p + 1)]^T. \tag{3.27}$$

Because $s(k)$ can be $+1$ or -1, there are $n_s = 2^{n+p}$ combinations of the channel input sequence.

We use the following nonlinear time-variant channel model:

$$r(k) = a_1 s(k) + a_2 s(k - 1) - 0.9[a_1 s(k) + a_2 s(k - 1)]^3 + e(k) \tag{3.28}$$

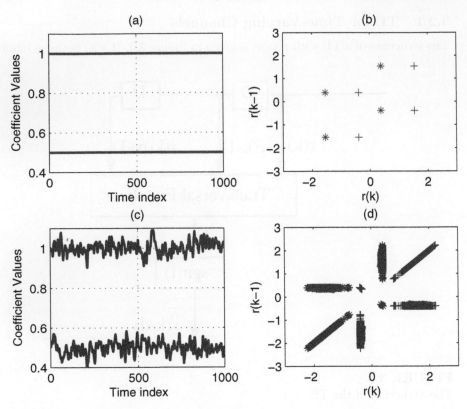

FIGURE 3.4 (See color insert.)
Modeling of a Nonlinear Time-Variant Channel: (a) Normal Channel Coefficient values a_i versus time, (b) Scattergram of received symbols, $r(k-1)$ and $r(k)$, (c) Channel coefficient values a_i versus time, in the presence of noise, and (d) Scattergram of received Symbols, $r(k-1)$ and $r(k)$ in the presence of noise.

where $a_1 = 1$ and $a_2 = 0.5$, as shown in Figure 3.4(a). We illustrate the design of a type-2 FAF for this channel, when the channel is time varying, i.e., when a_1 and a_2 are time-varying coefficients, each simulated using a second-order Markov model in which a Gaussian noise source drives a second-order Butterworth lowpass filter (LPF) (Liang and Mendel 2000). Note that we center $a_1(k)$ about 1 and $a_2(k)$ about 0.5, as shown in Figure 3.4(a).

The MATLAB code used in the above simulation is appended below:

```
%%% MATLAB script to simulate a Type-2 FAF..
%%% Last modified on 24-10-2012.
clear all;close all;clf;
a1=ones(1,1001);
a2=0.5*ones(1,1001);
```

```
subplot(221),plot([0:1000],a1,'LineWidth',2);
hold on;plot([0:1000],a2,'LineWidth',2);hold off;
axis([0 1000 0.4 1.1]);grid;
xlabel('Time index');ylabel('Coefficient Values');
title('(a)');
a1=1;a2=0.5;
sk=[1,1,1,1,-1,-1,-1,-1];
sk1=[1,1,-1,-1,1,1,-1,-1];
sk2=[1,-1,1,-1,1,-1,1,-1];
for k=1:4
    r(k)=a1*sk(k)+a2*sk1(k)-0.9*(sk(k)*a1+sk1(k)*a2)^3;
    r1(k)=a1*sk1(k)+a2*sk2(k)-0.9*(sk1(k)*a1+sk2(k)*a2)^3;
    subplot(222),plot(r,r1,'*');xlabel('r(k)');
    ylabel('r(k-1)');axis([-3 3 -3 3]);hold on;
    grid;
end;
for k=5:8
    r(k)=a1*sk(k)+a2*sk1(k)-0.9*(sk(k)*a1+sk1(k)*a2)^3;
    r1(k)=a1*sk1(k)+a2*sk2(k)-0.9*(sk1(k)*a1+sk2(k)*a2)^3;
    subplot(222),plot(r,r1,'+');grid;title('(b)');
end;
hold off;
[B,A]=butter(2,0.1);
beta=0.1;
a1=1+filter(B,A,beta*randn(1,1000));
a2=0.5+filter(B,A,beta*randn(1,1000));
subplot(223),plot([1:1000],a1,'LineWidth',2);
hold on; plot([1:1000],a2,'LineWidth',2);
hold off;xlabel('Time index');
ylabel('Coefficient Values');axis([0 1000 0.4 1.1]);grid;
title('(c)');
for k=1:4
    r=a1*sk(k)+a2*sk1(k)-0.9*(sk(k)*a1+sk1(k)*a2).^3;
    r1=a1*sk1(k)+a2*sk2(k)-0.9*(sk1(k)*a1+sk2(k)*a2).^3;
subplot(224),plot(r,r1,'*');hold on;axis([-3 3 -3 3]);
grid;
end;
for k=5:8
    r=a1*sk(k)+a2*sk1(k)-0.9*(sk(k)*a1+sk1(k)*a2).^3;
    r1=a1*sk1(k)+a2*sk2(k)-0.9*(sk1(k)*a1+sk2(k)*a2).^3;
subplot(224),plot(r,r1,'+');hold on;axis([-3 3 -3 3]);
grid;title('(d)');
end;
xlabel('r(k)'); ylabel('r(k-1)');hold off;
%%% end of fafin.m
```

TABLE 3.1
Channel States for Time-Varying Channel Model-I

$s(k)$	$s(k\text{-}1)$	$s(k\text{-}2)$	$\hat{r}(k)$
-1	-1	-1	$-a_1(k) - a_2(k) - 0.9[-a_1(k) - a_2(k)]^3$
-1	-1	1	$-a_1(k) - a_2(k) - 0.9[-a_1(k) - a_2(k)]^3$
-1	1	-1	$-a_1(k) + a_2(k) - 0.9[-a_1(k) + a_2(k)]^3$
-1	1	1	$-a_1(k) + a_2(k) - 0.9[-a_1(k) + a_2(k)]^3$
1	-1	-1	$a_1(k) - a_2(k) - 0.9[a_1(k) - a_2(k)]^3$
1	-1	1	$a_1(k) - a_2(k) - 0.9[a_1(k) - a_2(k)]^3$
1	1	-1	$a_1(k) + a_2(k) - 0.9[a_1(k) + a_2(k)]^3$
1	1	1	$a_1(k) + a_2(k) - 0.9[a_1(k) + a_2(k)]^3$

$s(k)$	$s(k\text{-}1)$	$s(k\text{-}2)$	$\hat{r}(k-1)$
-1	-1	-1	$-a_1(k) - a_2(k) - 0.9[-a_1(k) - a_2(k)]^3$
-1	-1	1	$-a_1(k) + a_2(k) - 0.9[-a_1(k) + a_2(k)]^3$
-1	1	-1	$a_1(k) - a_2(k) - 0.9[a_1(k) - a_2(k)]^3$
-1	1	1	$a_1(k) + a_2(k) - 0.9[a_1(k) + a_2(k)]^3$
1	-1	-1	$-a_1(k) - a_2(k) - 0.9[-a_1(k) - a_2(k)]^3$
1	-1	1	$-a_1(k) + a_2(k) - 0.9[-a_1(k) + a_2(k)]^3$
1	1	-1	$a_1(k) - a_2(k) - 0.9[a_1(k) - a_2(k)]^3$
1	1	1	$a_1(k) + a_2(k) - 0.9[a_1(k) + a_2(k)]^3$

Observe that the channel states are now eight clusters instead of eight individual points, as shown in Figure 3.4(d). From Table 3.1, we see that there are eight channel states and that $s(k)$ determines which cluster $\hat{r}(k)$ belongs to.

3.2.1.1 Designing the Type-2 FAF

In our type-2 FAF design, there are eight rules (each rule corresponds to one channel state), where the lth rule R^l is expressed as

$$R^l : IF \ r(k) \ is \ \tilde{F}_1^l \ and \ r(k-1) \ is \ \tilde{F}_2^l \ THEN \ y^l = w_l$$

where $\tilde{F}_1^l$ and $\tilde{F}_2^l$ are type-2 Gaussian MFs with uncertain means, and w_l is a crisp value of $+1$ or -1 as determined by $\hat{r}(k)$. For rule l, the range of the mean of antecedent $\tilde{F}_1^l$ ($\tilde{F}_2^l$) corresponds to the horizontal (vertical) projection of the lth cluster in Figure 3.4(d). Observe from this rule that the consequent is a constant (i.e., it does not depend on $r(k)$ and $r(k-1)$).

Equation 3.24 is used to compute the output of the type-2 FAF, where $y^l = w_l$ ($l = 1, \ldots, 8$) equals 1 or -1, $\underline{f}^l$ is obtained from Equation 3.19 and $\overline{f}^l$ is obtained from Equation 3.20. We chose

$$\mu_{\tilde{F}_k^l}(x_k) = exp\left[-\frac{1}{2}\left(\frac{x_k - m_k^l}{\sigma_e}\right)^2\right], \tag{3.29}$$

$$where \ m_k^l \in \left[m_{k1}^l, m_{k2}^l\right] \tag{3.30}$$

and $k = 1, 2$.

The MATLAB® code to generate the observation space in Figure 3.11 is appended below:

```
%%% MATLAB script to simulate the Observation Space..
%%% for a Radio Channel..
%%% Last modified on 24-10-2012.
clc;clear all;close all;clf;
sk=[-1,-1,-1,-1,1,1,1,1,-1,-1,-1,-1,1,1,1,1];
sk1=[-1,-1,-1,-1,-1,-1,-1,-1,1,1,1,1,1,1,1,1];
sk2=[-1,-1,1,1,-1,-1,1,1,-1,-1,1,1,-1,-1,1,1];
sk3=[-1,1,-1,1,-1,1,-1,1,-1,1,-1,1,-1,1,-1,1];
a1=0.3482;a2=0.8704;a3=0.3482;
for k=1:8
    r(k)=a1*sk(k)+a2*sk1(k)+a3*sk2(k);
    r1(k)=a1*sk1(k)+a2*sk2(k)+a3*sk3(k);
    plot(r,r1,'*');xlabel('r(k)'); ylabel('r(k-1)');
            axis([-3 3 -3 3]);hold on;
end;
for k=9:16
    r(k)=a1*sk(k)+a2*sk1(k)+a3*sk2(k);
    r1(k)=a1*sk1(k)+a2*sk2(k)+a3*sk3(k);
```

```
   plot(r,r1,'o');
end;
hold off;grid;
title('Observation Space for Radio Channel');
%%% end of af.m..
```

To complete the specification of the membership functions in 3.30, we also need to estimate the standard deviation of the noise, σ_e. It can be shown that equalizer performance is not very sensitive to the value of σ_e. In the simulations, we assume that the value of σ_e is known exactly. We fix SNR and compute the standard deviation $\sigma_{\hat{r}}$ of $\hat{r}(k)$ in the combined training and testing sequence. Then based on the fact that

$$SNR = 10 \log_{10} \frac{\sigma_{\hat{r}}^2}{\sigma_e^2} \tag{3.31}$$

we compute σ_e as

$$\sigma_e = \sigma_{\hat{r}}/10^{\frac{SNR}{20}}. \tag{3.32}$$

3.2.1.2 Simulations

We compare the type-2 FAF with an unnormalized type-1 FAF (the latter is identical to an RBF network) and a Nearest Neighbor Classifier (NNC) (Savazzi 1998) for equalization of the time-varying channel given in Equation 3.28. In the simulations, the number of taps of the equalizer p is made equal to the number of taps of the channel, $n+1$; i.e., $p = n+1$. The number of rules is equal to the number of clusters, i.e., 2^{p+n}. A sequence $s(k)$ of length 1000 is used in the simulations. The number of training prototypes is chosen as 121 (the first 121 symbols), and the remaining 879 are used for testing. This is due to the fact that the number of training prototypes should be a *perfect square* in the case of an NNC.

In the first trial, we fix SNR at 20 dB and run simulations for eight different values of β (the standard deviation of the white Gaussian noise sequence used to generate tap-coefficients) from 0.04 to 0.32 with step size 0.04 ([0.04:0.04:0.32]) and we set the decision delay, $d = 0$. We perform 100 Monte Carlo simulations for each β value and plot the mean and standard deviations of the BER. Figure 3.5 shows the plots.

In the second trial, we fix $\beta = 0.1$ and run simulations for five different SNR values ranging from 15 to 25 dB ([15:2.5:25]). We again perform 100 Monte Carlo simulations for each SNR value. The results are plotted as shown in Figure 3.6.

3.2.1.3 Observations

The following observations are made from the above figures.

1. In terms of mean values of BER, the type-2 FAF performs much better than both the NNC and the type-1 FAF.

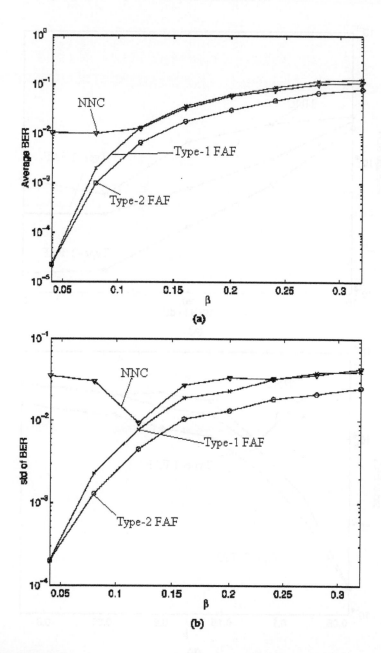

FIGURE 3.5
Performance of TE-I: (a) Average BER versus Standard Deviation of AWGN; Average BER varies from 10^{-5} to 10^{0}, standard deviation of AWGN varies from 0.05 to 0.35, (b) Standard deviation of BER (varies from 10^{-4} to 10^{-1}) versus standard deviation of AWGN (varies from 0.05 to 0.35).

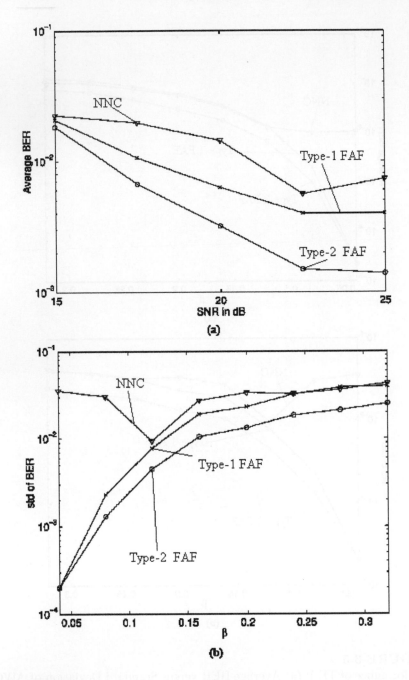

FIGURE 3.6
Performance of TE-II: (a) Average BER (varies from 10^{-3} to 10^{-1}) versus
SNR (varies from 15 to 25), and (b) Standard deviation of BER (varies from
10^{-4} to 10^{-1}) versus standard deviation of AWGN (varies from 0.05 to 0.35).

2. When the SNR is 20 dB, the NNC performs better than the type-1 FAF when $\beta \geq 0.12$ and the type-1 FAF performs better than the NNC when $\beta < 0.12$, but the type-2 FAF always performs better than the NNC.

3. In terms of standard deviation of BER, the type-2 FAF is more robust to the additive Gaussian noise than the other two equalizers (Liang and Mendel 2000).

These observations suggest that a type-2 FAF holds promise as a good TE for time-varying channels. Unfortunately, though, the number of rules for such an equalizer is $n_s = 2^{n+p}$ (recall that $n+1$ is the number of channel taps and p is the number of antecedents). For example, for $n = 4$ and $p = 5$, we need 512 rules. This causes huge computational complexity when the channel order is high.

3.2.2 DFE for Time-Varying Channel Using a Type-2 FAF

It is well known that a DFE can reduce computational complexity and improve equalization performance as compared to a TE. Figure 3.7 shows the structure of a DFE having p feedforward taps and q feedback taps (Liang and Mendel 2000).

3.2.2.1 Design of a DFE Based on a Type-2 FAF

We use the following nonlinear time-varying channel in the simulations of a DFE:

$$
\begin{aligned}
r(k) &= a_1(k)s(k) + a_2(k)s(k-1) + a_3(k)s(k-2) \\
&\quad -0.7[a_1(k)s(k) + a_2(k)s(k-1) + a_3(k)s(k-2)]^3 + e(k) \quad (3.33)
\end{aligned}
$$

where nominal values for the channel coefficients are $a_1 = 0.3482$, $a_2 = 0.8704$, $a_3 = 0.3482$. A nonlinear channel model like 3.33 is frequently encountered in data transmission over satellite links, especially when the signal amplifiers operate in their high gain limits, which results in nonlinearity.

The decision delay d is assumed to be 1. Since the channel order $n = 2$, it is sufficient to design a DFE with two feedforward taps (i.e., $p = 2$) and two feedback taps (i.e., $q = 2$), which means that the *decision tree* has $2^2 = 4$ leaves (FAFs), and each leaf (FAF) has $2^2 = 4$ rules.[2] The channel states of Equation 3.33 are given in Table 3.2. The architecture of a DFE based on FAFs is shown in Figure 3.8. Designing rules in each of the four FAFs are the same as that of designing a transversal fuzzy equalizer. There are a total of 16 rules and the lth rule, R^l, is expressed as

$$
R^l : \quad IF \ r(k) \ is \ \tilde{F}_1^l \ and \ r(k-1) \ is \ \tilde{F}_2^l \ THEN \ y^l = w^l
$$

---

[2]The *decision tree* refers to the paths each estimated symbol $\hat{s}(k)$ can assume at the detector side.

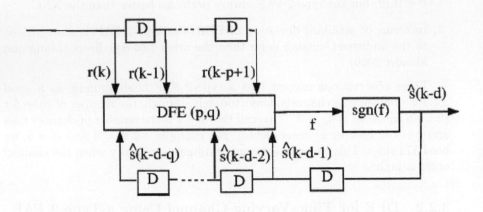

FIGURE 3.7
Structure of the DFE.

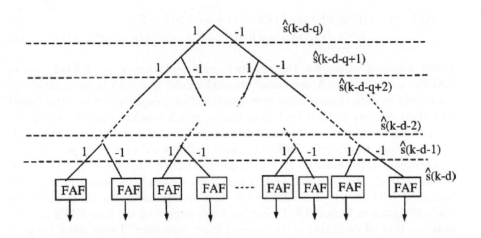

FIGURE 3.8
Architecture of the DFE Based on FAF.

TABLE 3.2

Channel States for Time-Varying Channel Model-II

$s(k)$ $s(k-1)$ $s(k-2)$ $s(k-3)$	$\hat{r}(k)$	$\hat{r}(k-1)$
(1 1 1 1)	$a_1(k) + a_2(k) + a_3(k)$	$a_1(k) + a_2(k) + a_3(k)$
(-1 1 1 1)	$-a_1(k) + a_2(k) + a_3(k)$	$a_1(k) + a_2(k) + a_3(k)$
(1 -1 1 1)	$a_1(k) - a_2(k) + a_3(k)$	$-a_1(k) + a_2(k) + a_3(k)$
(-1 -1 1 1)	$-a_1(k) - a_2(k) + a_3(k)$	$-a_1(k) + a_2(k) + a_3(k)$
(1 1 1 -1)	$a_1(k) + a_2(k) + a_3(k)$	$a_1(k) + a_2(k) - a_3(k)$
(-1 1 1 -1)	$-a_1(k) + a_2(k) + a_3(k)$	$a_1(k) + a_2(k) - a_3(k)$
(1 -1 1 -1)	$a_1(k) - a_2(k) + a_3(k)$	$-a_1(k) + a_2(k) - a_3(k)$
(-1 -1 1 -1)	$-a_1(k) - a_2(k) + a_3(k)$	$-a_1(k) + a_2(k) - a_3(k)$
(1 1 -1 1)	$a_1(k) + a_2(k) - a_3(k)$	$a_1(k) - a_2(k) + a_3(k)$
(-1 1 -1 1)	$-a_1(k) + a_2(k) - a_3(k)$	$a_1(k) - a_2(k) + a_3(k)$
(1 -1 -1 1)	$a_1(k) - a_2(k) - a_3(k)$	$-a_1(k) - a_2(k) + a_3(k)$
(-1 -1 -1 1)	$-a_1(k) - a_2(k) - a_3(k)$	$-a_1(k) - a_2(k) + a_3(k)$
(1 1 -1 -1)	$a_1(k) + a_2(k) - a_3(k)$	$a_1(k) - a_2(k) - a_3(k)$
(-1 1 -1 -1)	$-a_1(k) + a_2(k) - a_3(k)$	$a_1(k) - a_2(k) - a_3(k)$
(1 -1 -1 -1)	$a_1(k) - a_2(k) - a_3(k)$	$-a_1(k) - a_2(k) - a_3(k)$
(-1 -1 -1 -1)	$-a_1(k) - a_2(k) - a_3(k)$	$-a_1(k) - a_2(k) - a_3(k)$

where we assume that $\tilde{F}_1^l$ and $\tilde{F}_2^l$ are type-2 Gaussian membership functions with uncertain means, and w_l is a crisp value of $+1$ or -1. The algorithm for designing a FAF DFE based on N training prototypes for a channel with $n + 1$ taps and a decision delay of d (which determines $p = d + 1$ and $q = n$) is as follows:

1. Based on the values of $[s(k - d - q), \ldots, s(k - d - 1)]$, a branch and its corresponding leaf (FAF) in the decision tree is chosen.

2. In the chosen FAF, design 2^p rules, which means 2^p clusters are needed. Based on $[s(k - d), \ldots, s(k)]$, we know which cluster $\mathbf{r}(k)$ belongs to and $s(k - d)$ determines the cluster category $+1$ or -1.

3. Repeat steps 1 and 2 until all N training prototypes have been clustered.

4. Suppose in the ith FAF, $(i = 1, \ldots, 2^p)$, there are N_l training prototypes belonging to the lth cluster, $(l = 1, \ldots, 2^p)$, and the mean and standard deviation of these are $\mathbf{r}(k)$ $[p \times 1\ \ vector]$ and σ_r^l $[p \times 1\ \ vector]$, respectively. Now obtain the parameters $\mathbf{m_1^l}$ and $\mathbf{m_2^l}$, where

$$\mathbf{m_1^l} = [m_{11}^l, m_{21}^l, \ldots, m_{p1}^l]^T$$
$$\mathbf{m_2^l} = [m_{12}^l, m_{22}^l, \ldots, m_{p2}^l]^T$$

so $[m_{j1}^l, m_{j2}^l]$ $(j = 1, 2, \ldots, p)$ is the range of the type-2 antecedent Gaussian membership function, $\mu_{\tilde{F}_j^l}$, in the ith FAF.

5. After the training period, the parameters of every FAF are fixed. In the testing period, for every $\mathbf{r(k)}$, use results given in Equations 3.19, 3.20, and 3.24 to compute the defuzzified output of the activated FAF $f(x)$ and then obtain the output of the DFE $\hat{s}(k-d)$.

3.2.2.2 Simulations

Simulations are performed for channel (3.33) in which a 1000 symbol sequence $s(k)$ is used. The first 289 symbols are used for training and the remaining 711 symbols are used for testing. After training, the parameters in all four fuzzy filters are fixed and then testing is performed.

In the first trial, the SNR is fixed at 20 dB and the simulations are run for five different β (the standard deviation of the white Gaussian noise sequence) ranging from 0.04 to 0.2 ([0.04:0.04:0.2]). One hundred Monte Carlo simulations are performed for each β value and the results are plotted as indicated in Figure 3.9.

In a second trial, the β is fixed at 0.1 and the simulations are run for seven different SNRs ranging from 15 to 30 dB ([15:2.5:30]). Figure 3.10 shows the results.

3.2.2.3 Observations

From the mean and standard deviations of the BER, we conclude that the DFE based on four type-2 FAFs performs much better than the NNC and the DFE based on four type-1 FAFs (each is an unnormalized type-1 TSK FLS).

3.2.3 Inferences

We use a type-2 FAF to implement a TE and also use a decision tree and more than one type-2 FAF to implement a DFE. The following conclusions can be made after exhaustive simulations:

1. The number of rules in each FAF of the DFE is tremendously reduced. For a channel with $n+1$ taps, the rule reduction ratio is 2^n:1.

2. Both the type-2 FAF TE and DFE perform better than either a type-1 FAF or an NNC.

3. Since no tuning procedure is used in the design of either type-2 FAF based equalizer, real-time information processing is guaranteed.

4. It will be of great advantage to develop a FAF-based *Blind Channel Equalizer*.

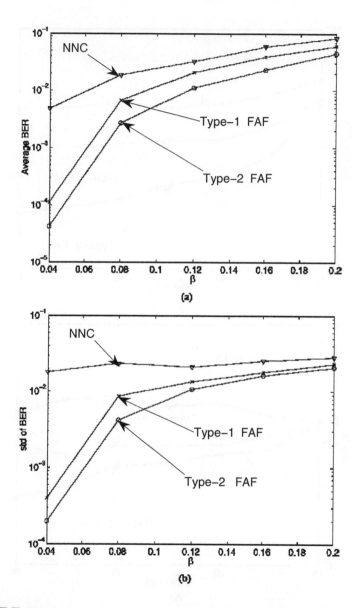

FIGURE 3.9
Performance of DFE Based on FAF-I: (a) Average BER (varies from 10^{-5} to 10^{-1}) versus Standard Deviation of AWGN (varies from 0.04 to 0.2), and (b) Standard deviation of BER (varies from 10^{-4} to 10^{-1}) versus standard deviation of AWGN (varies from 0.04 to 0.2).

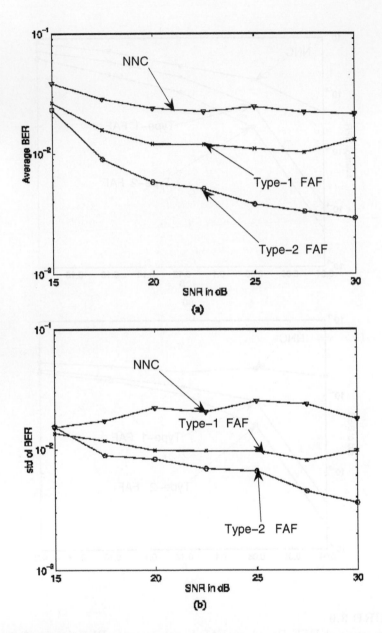

FIGURE 3.10
Performance of DFE Based on FAF-II: (a) Average BER (varies from 10^{-3} to 10^{-1}) versus SNR in dB (varies from 15 to 30), and (b) Standard deviation of BER (varies from 10^{-3} to 10^{-1}) versus SNR in dB (varies from 15 to 30).

3.3 Adaptation of the Type-2 FAF for the Indoor Environment

Before we make type-2 FAF based DFE for the Mobile Cellular Indoor channel, we have to choose the most suitable channel model for the above environment. In continuation of the discussion in the first part of Chapter-2, we will now consider certain indoor propagation models. With the advent of Personal Communication Systems (PCS), there is a great deal of interest in characterizing radio propagation inside buildings (Rappaport 2003). The indoor radio channel differs from the traditional radio channel in two aspects–the distances covered are much smaller, and the variability of the environment is much greater for a much smaller range of T–R separation distances. The article by Hashemi discusses indoor radio propagation models (Hashemi 1993). Some of the important models which have recently emerged are presented below.

3.3.1 Log–Distance Path Loss Model

Indoor path loss obeys the distance power law in Equation 3.34:

$$PL(dB) = PL(d_0) + 10n \log \left[\frac{d}{d_0} \right] + X_\sigma \qquad (3.34)$$

where the value of n depends on the surroundings and building type, and X_σ represents a normal random variable in dB having a standard deviation of σ dB (Rappaport 2003).

3.3.2 Ericsson Multiple Breakpoint Model

The Ericsson radio system model is obtained by measurements in a multiple floor office building. The model has four breakpoints and considers both an upper and lower bound on the path loss. The model also assumes that there is 30 dB attenuation at $d_0 = 1$ m, which can be shown to be accurate for $f = 900$ MHz and unity gain antennae (Rappaport 2003).

3.3.3 Attenuation Factor Model

An in-building site-specific propagation model that includes the effect of building type as well as the variations caused by obstacles is described by Seidel (Seidel and Rappaport 1992) and has been used to accurately deploy indoor and campus networks. The attenuation factor model is given by

$$\overline{PL}(d)[dB] = \overline{PL}(d_0)[dB] + 10n_{SF} \log \left[\frac{d}{d_0} \right] + \mathcal{FAF}[dB] + \sum \mathcal{PAF}[dB]$$
$$(3.35)$$

where n_{SF} represents the exponent value for the "same floor" measurement, $\mathcal{FAF}$ represents a *floor attenuation factor* for a specified number of building floors, and $\mathcal{PAF}$ represents the partition attenuation factor for a specific obstruction encountered by a ray drawn between the transmitter and receiver in 3-D. Alternatively, in Equation 3.35, $\mathcal{FAF}$ may be replaced by an exponent which already considers the effects of multiple floor separation (Rappaport 2003).

$$\overline{PL}(d)[dB] = \overline{PL}(d_0)[dB] + 10n_{MF} \log\left[\frac{d}{d_0}\right] + \sum \mathcal{PAF}[dB] \qquad (3.36)$$

where n_{MF} denotes a path loss exponent based on measurements through multiple floors.

Devasirvatham et al. find that in-building path loss obeys free space plus an additional loss factor which increases exponentially with distance. Based on this, it will be possible to modify Equation 3.35 such that

$$\overline{PL}(d)[dB] = \overline{PL}(d_0)[dB] + 20 \log\left[\frac{d}{d_0}\right] + \alpha\, d + \mathcal{FAF}[dB] + \sum \mathcal{PAF}[dB]$$
$$(3.37)$$

where α is the attenuation constant for the channel with units of dB per meter (dB/m) (Rappaport 2003).

3.3.4   DFE for an Indoor Mobile Radio Channel

We now consider a DFE for the indoor channel based on the RBF neural network.

3.3.4.1   Channel Equation

We use the following equation for the Channel Impulse Response (CIR):

$$h_{ch1} \;=\; 0.3482\,\delta(n) + 0.8704\,\delta(n-1) + 0.3482\,\delta(n-2). \qquad (3.38)$$

For the DFE the vector of received samples $\hat{r}(k)$ contains two samples, i.e., $M = 2$; the feedback vector $\mathbf{s_D}(k)$ contains 2 elements, i.e., $D = 2$; the estimation lag, $d = 1$. Proceeding in a similar manner to the example of Table 3.2, we construct a state table (Table 3.3) that relates the input signal states of channel input vector $\hat{r}(k)$ to $\hat{s}(k)$ in the absence of noise (Mulgrew 1996). The states have been numbered to aid interpretation.

At a particular sample, k, the contents of $\mathbf{s_D}$ reduce the number of possible output states of the vector $r(k)$ from 16 to 4. For example, if

$$\mathbf{s_D}(\mathbf{k}) = [\hat{s}(k-2)\ \hat{s}(k-3)]^T = [1\ 1]^T$$

then only states 3, 7, 11, or 15 could have been received, and hence only the centers associated with these states will be used in the Bayesian or RBF

TABLE 3.3

Channel States for Time-Varying Channel Model-III

#	$s(k)$	$s(k\text{-}1)$	$s(k\text{-}2)$	$s(k\text{-}3)$	$\hat{r}(k)$	$\hat{r}(k-1)$
0	-1	-1	-1	-1	-1.57	-1.57
1	-1	-1	-1	1	-1.57	-0.87
2	-1	-1	1	-1	-0.87	0.17
3	-1	1	1	-1	-0.87	0.87
8	1	-1	-1	-1	-0.87	-1.57
9	1	-1	-1	1	-0.87	-0.87
10	1	-1	1	-1	-0.17	0.17
11	1	-1	1	1	-0.17	0.87
4	-1	1	-1	-1	0.17	-0.87
5	-1	1	-1	1	0.17	-0.17
6	-1	1	1	-1	0.87	0.87
7	-1	1	1	1	0.87	1.57
12	1	1	-1	-1	0.87	-0.87
13	1	1	-1	1	0.87	-0.17
14	1	1	1	-1	1.57	0.87
15	1	1	1	1	1.57	1.57

network. Thus, the role for the vector $\mathbf{s_D}$ in the decision feedback structure is to select a subset of centers for a particular decision, rather than providing additional terms for the vector input to a neural network.

The superior performance of the DFE structure in comparison with the feedforward structure is suggested in Figure 3.11.

Decision errors are a function of the distance of the centers to the decision boundary. The further the centers are from the boundary, the lower the probability of misclassification. States 5 and 10 are the closest centers to the feedforward boundary, and hence will heavily influence the probability of misclassification. The optimal boundary for the DFE is determined by centers 3, 7, 11, and 15 alone. Of these, 7 and 11 are the closest to the boundary. These states are farther from the feedforward boundary than 5 and 10 are, which suggests that the performance of the DFE will be superior to the feedforward equalizer.

In addition to improving the performance of the equalizer decision, feedback reduces the complexity, in that at any time period only a subset of the basis functions is used to form the decision function. To further improve the computational complexity, a technique originally proposed in Clark et al. (1982) and extended in the block Bayesian equalizer of Williamson et al. (1992) can be used. Consider the example of Table 3.3. Again assume that the decisions $\hat{r}(k-2)$ and $\hat{r}(k-3)$ are correct. The state equation that relates

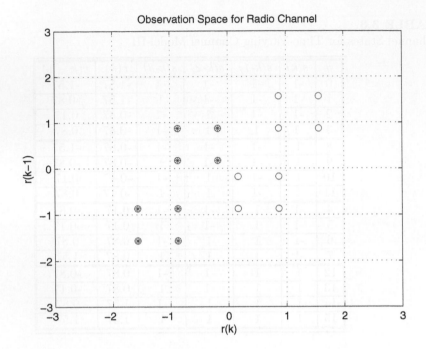

FIGURE 3.11 (See color insert.)
Observation Space for Radio Channel: Scattergram showing $r(k-1)$ versus $r(k)$, the received symbols at adjoining time intervals. Bullets indicate one *class* of states, whereas circles indicate another.

the received signal vector $\mathbf{r}(\mathbf{k})$ to the vector of transmitted symbols is

$$\mathbf{r}(\mathbf{k}) = \mathbf{H}\,\mathbf{s}(\mathbf{k}) + \mathbf{n}(\mathbf{k})$$

$$
\left[\begin{array}{c} r(k) \\ r(k-1) \end{array} \right]
=
\left[\begin{array}{cc|cc} h_0 & h_1 & h_2 & 0 \\ 0 & h_0 & h_1 & h_2 \end{array} \right]
\left[\begin{array}{c} s(k) \\ s(k-1) \\ -- \\ \hat{s}(k-2) \\ \hat{s}(k-3) \end{array} \right]
+
\left[\begin{array}{c} n(k) \\ n(k-1) \end{array} \right]
$$

$$(3.39)$$

The partitioning of $\mathbf{H}$ and $\mathbf{s}(\mathbf{k})$ highlights the contribution of previous decisions on the observed vector $\mathbf{r}(\mathbf{k})$.

In this example, the feedforward equalizer $s(k)$ has 4 elements, and hence 2^4 centers are required, whereas in the DFE, $s_1(k)$ has 2 elements and hence 2^2 centers are required. The decision function is implemented using a radial basis function network. For this form of equalizer, direct estimation of the channel impulse response is required to form the feedback matrix $\mathbf{H_2}$ and hence this estimate is also used to calculate the centers for the RBF. The weights in the output are assigned at the end of the training period.

3.3.5 Co-Channel Interference Suppression

Although the primary reason for using an equalizer on a communications channel is to mitigate the effects of intersymbol interference, more recently it has been demonstrated that conventional equalizers can exploit the cyclostationary nature of the received signal and reduce the distortion due to both co-channel and adjacent channel interference. A radial basis function network can also be applied to this problem without the need to exploit the cyclostationary characteristics of the received signal (Mulgrew 1996).

The received signal vector $\mathbf{r}(k)$ is now composed of three rather than two terms: an own-channel component $\mathbf{H}\,\mathbf{s}(k)$ and a similar co-channel component $\mathbf{H_c}\,\mathbf{s_c}(k)$, and a noise term, $n(k)$. Thus:

$$\mathbf{r}(k) = [\mathbf{H} \mid \mathbf{H_c}] \begin{bmatrix} \mathbf{s}(k) \\ -- \\ \mathbf{s_c}(k) \end{bmatrix} + n(k) \qquad (3.40)$$

The effect of the co-channel is to increase the number of centers shown in Figure 3.11, because the aggregate vector $[\mathbf{x}^T(k)\,\mathbf{x_c}^T(k)]^T$ has by definition more elements than $\mathbf{x}(k)$. By using a two stage training process of supervised and unsupervised clustering, optimal performance can be attained (Mulgrew 1996).

3.4 Conclusion

In earlier sections of this chapter, a DFE based on a type-2 FAF is considered. It is shown that it outperforms the type-1 FAF and NNC. In the latter sections of the chapter, we adapt the Bayesian DFE for the indoor mobile channel. We arrive at the following important conclusions:

1. The adaptive Bayesian DFE provides a useful alternative to both linear and maximum likelihood equalizers in terms of the complexity/performance trade-off.

2. The number of rules required is less and consequently the training time required is also less. It is seen that for a channel with $n+1$ taps, the rule reduction ratio is $2^n{:}1$.

3. The application of RBF and adaptive Bayesian methods to CDMA systems is at a much earlier stage than its ISI counterpart. Current results indicate that they do provide performance somewhere between linear and maximum likelihood methods.

4. Although FAF has been extensively used for channel equalization, a training sequence is needed for all approaches. To develop a *blind FAF equalizer* is a challenging problem.

5. The use of the RBF has provided receivers with more controllable training characteristics than Multi-Level Perceptron (MLP) receivers. However, the length of the training period is still too long for practical consideration.

6. In terms of complexity, the Bayesian DFE is more expensive than conventional DFE solutions and less expensive than standard Maximum Likelihood Viterbi Algorithm (MLVA) solutions.

Further Reading

Qilian Liang and Jerry M. Mendel, Equalization of Nonlinear Time-Varying Channels Using Type2 Fuzzy Adaptive Filters, *IEEE Transactions on Fuzzy Systems*, Vol.8, No.5, pp.551–563, October 2000.

Tulay Adali, Why a Nonlinear Solution for a Linear Problem? *Proceedings of IEEE International Workshop on Neural Networks for Signal Processing*, pp. 157–165, 1999.

Nan Xie and Henry Leung, Blind Equalization Using a Predictive Radial Basis Function Neural Network, *IEEE Transactions on Neural Networks*, Vol.16, No.3, pp.709–720, May 2005.

Chin-Teng Lin and C.S. George Lee, *Neural Fuzzy Systems*, International Edition, Prentice Hall, New York 1996.

Lotfi A. Zadeh, Fuzzy Sets, *Information and Control*, Vol.8, pp.338–353, 1965.

N.N. Karnik, J.M. Mendel and Q. Liang, Type2 Fuzzy Logic Systems, *IEEE Transactions on Fuzzy Systems*, Vol.7, pp.643–658, December 1999.

Jerry M. Mendel, Uncertainty, Fuzzy Logic, and Signal Processing, *IEEE Signal Processing*, Vol.80, No.6, pp.913–933, June 2000.

Qilian Liang and Jerry M. Mendel, Overcoming Time-Varying Co-Channel Interference Using Type2 Fuzzy Adaptive Filters, *IEEE Transactions on Circuits and Systems II:Analog and Digital Signal Processing*, Vol.47,No.12, pp.1419–1428, December 2000.

Pietro Savazzi, Lorenzo Favalli, Eugenio Costamagna, and Alessandro Mecocci, A Suboptimal Approach to Channel Equalization Based on the Nearest Neighbor Rule, *IEEE Journal of Selected Areas in Communications*, Vol.16, No.9, pp.1640–1648, December 1998.

FIGURE 1

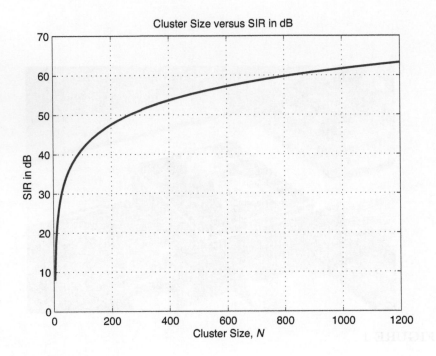

FIGURE 2.2
A Plot of Cluster Size Versus SIR in dB.

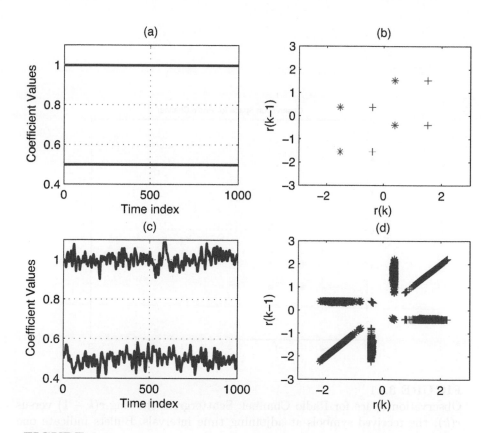

FIGURE 3.4

Modeling of a Nonlinear Time-Variant Channel: (a) Normal Channel Coefficient values a_i versus time, (b) Scattergram of received symbols, $r(k-1)$ and $r(k)$, (c) Channel coefficient values a_i versus time, in the presence of noise, and (d) Scattergram of received Symbols, $r(k-1)$ and $r(k)$ in the presence of noise.

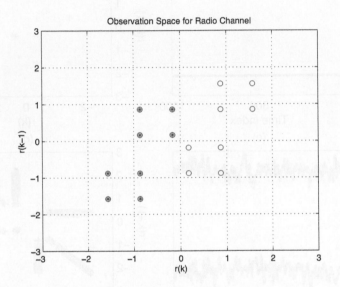

FIGURE 3.11

Observation Space for Radio Channel: Scattergram showing $r(k-1)$ versus $r(k)$, the received symbols at adjoining time intervals. Bullets indicate one *class* of states, whereas circles indicate another.

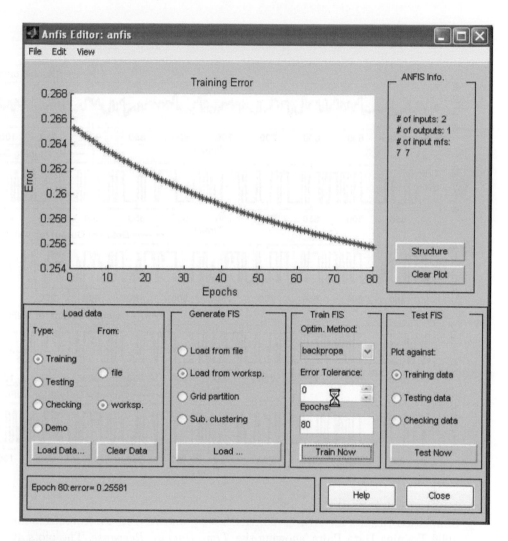

FIGURE 4.7
The Error Plot of Training of ANFIS–27; Generated Using MATLAB Fuzzy
Logic Toolbox: Number of Inputs = 2, Number of Outputs = 1, Total Number
of Fuzzy Rules = 49, Type of Membership Function: Gaussian, and Number
of Epochs = 80.

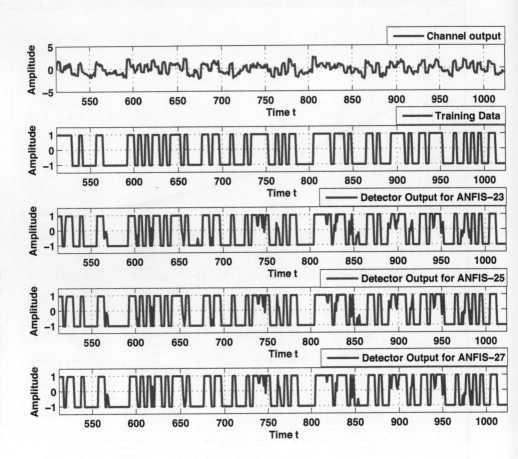

FIGURE 4.8
Simulation Results for ANFIS–23, ANFIS–25, and ANFIS–27 Equalizers for 4096 Training Data Pairs, showing the *Time Domain Response*. The plots at the bottom represent the output of ANFIS–23, ANFIS–25, and ANFIS–27 equalizers with an attached hard thresholding detector.

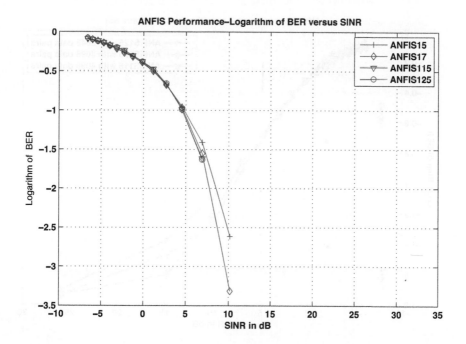

FIGURE 4.9
Performance of ANFIS Equalizers. Logarithm of BER at the output of the
equalizer versus SINR in dB (varies from −26 to −8).

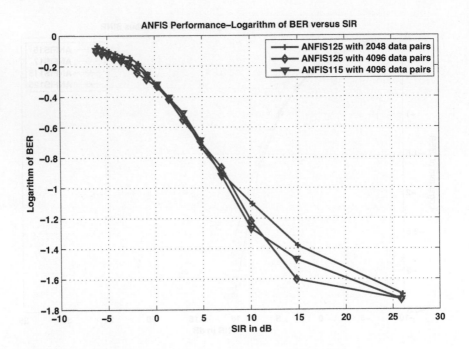

FIGURE 4.10
Performance of ANFIS Equalizers. Logarithm of BER at the output of the equalizer versus SIR in dB (varies from −10 to 30, with standard deviation of AWGN fixed at 0.42).

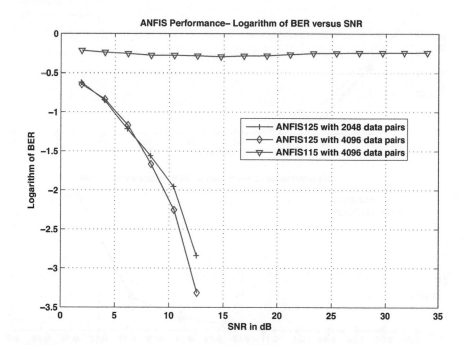

FIGURE 4.11

Simulation Results: Logarithm of BER at the output of the equalizer (varies from 0 to −3.5) versus SNR in dBs (varies from 0 to 35) for 2048/4096 training data pairs with standard deviation of CCI fixed at 0.08.

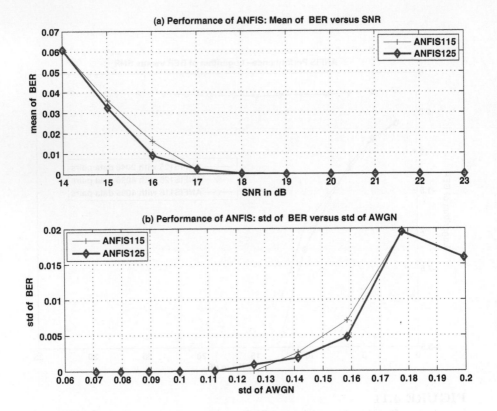

FIGURE 4.12

Performance of ANFIS Equalizers: (a) Mean BER at the output of the equalizer (varies from 0 to 0.07) versus SNR in dBs (varies from 14 to 23), and (b) standard deviation of BER at the output of the equalizer (varies from 0 to 0.02) versus standard deviation of AWGN (varies from 0.06 to 0.2).

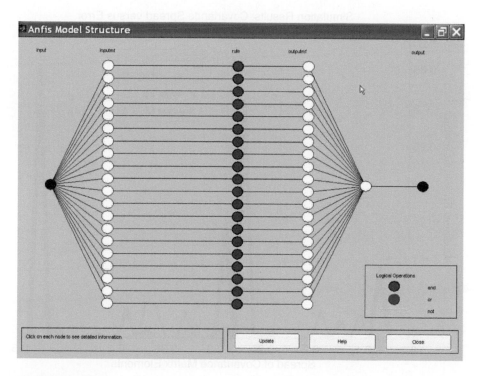

FIGURE 4.13
ANFIS Model Structure Used for UWB Channel Equalization: Number of Inputs=1, Number of Outputs=1, Number of Rules=20, and Type of Membership Function- Gaussian.

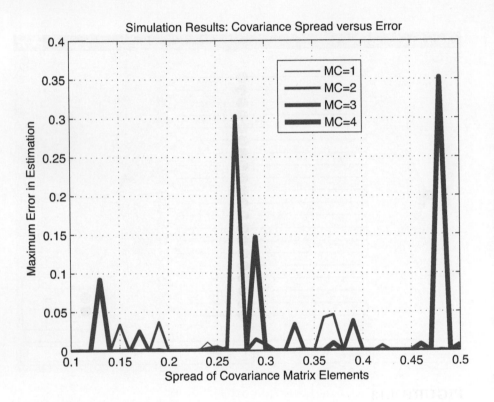

FIGURE 4.14
Results of Simulation–ANFIS Equalizer for UWB Channels: The results of varying number of trials are indicated by varying line thicknesses. Note that results improve as the number of trials increases.

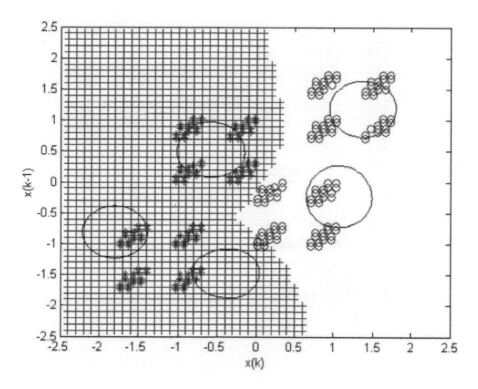

FIGURE 5.2
Decision Boundaries of the CNFF for $k = 50$. The boundary is marked by the mesh. The plot is a scattergram of the symbols received, $x(k-1)$ and $x(k)$, at consecutive instances.

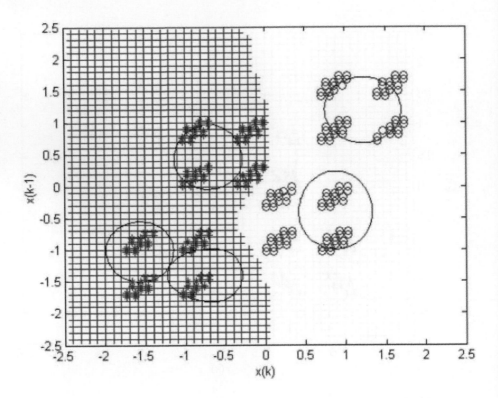

FIGURE 5.3

Decision Boundaries of the CNFF for $k = 100$. The boundary is marked by the mesh. The plot is a scattergram of the symbols received, $x(k-1)$ and $x(k)$, at consecutive instances.

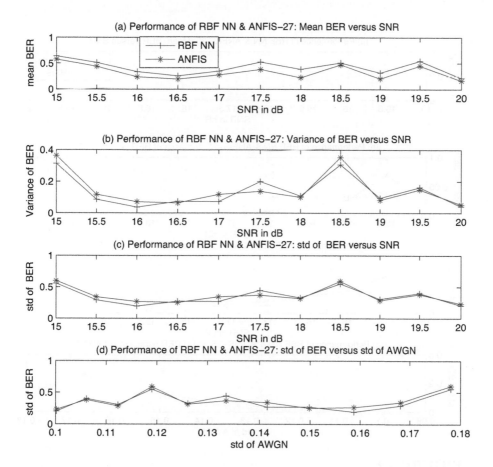

FIGURE 6.1

Performance of RBF NN and ANFIS-27: (a) Mean BER at output of the equalizer versus SNR in dB, (b) Variance of BER versus SNR in dB, (c) standard deviation of BER versus SNR in dBs, and (d) Standard deviation of BER versus standard deviation of AWGN.

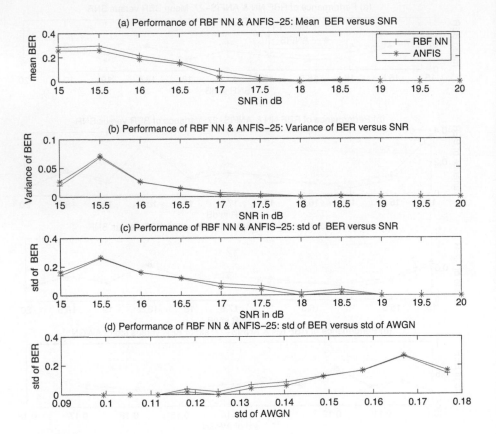

FIGURE 6.2
Performance of RBF NN and ANFIS-25: (a) Mean BER at output of the equalizer versus SNR in dB, (b) Variance of BER versus SNR in dB, (c) Standard deviation of BER versus SNR in dBs, and (d) Standard deviation of BER versus standard deviation of AWGN.

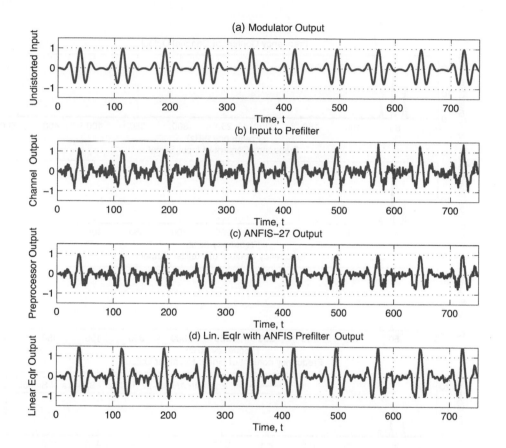

FIGURE 7.3

Simulation Results of Preprocessor Scheme with ANFIS Prefilter, Showing the *Time Domain Responses*. The input signal $x(t)$ is a Gaussian pulse train the output of the channel is given by $y(t) = x(t) + 0.2x^2(t) - 0.1x^3(t) + \eta(t)$. Note that waveforms (a) and (d) are closer to each other than (a) and (c).

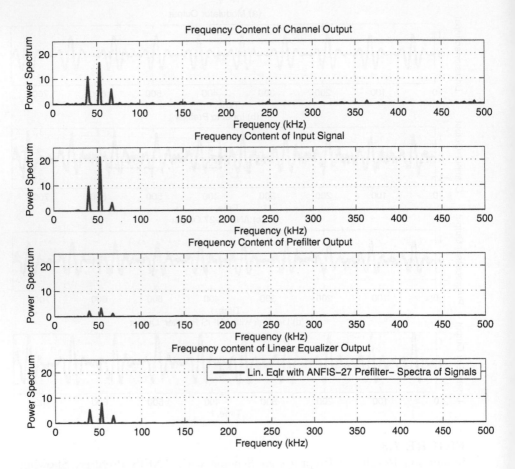

FIGURE 7.4

Spectra of Signals in Figure 7.3.

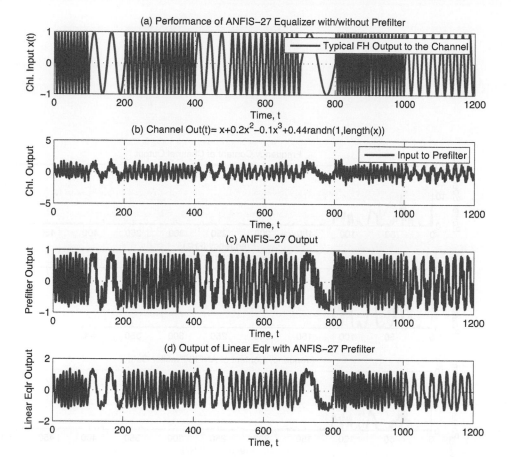

FIGURE 7.5

Simulation Results with ANFIS Prefilter for a Frequency-Hopped (FH) Carrier. The signal $x(t)$ is an FH carrier; the output $y(t) = x(t) + 0.2x^2(t) - 0.1x^3(t) + \eta(t)$. Note that waveforms (a) and (d) are closer to each other than (a) and (c).

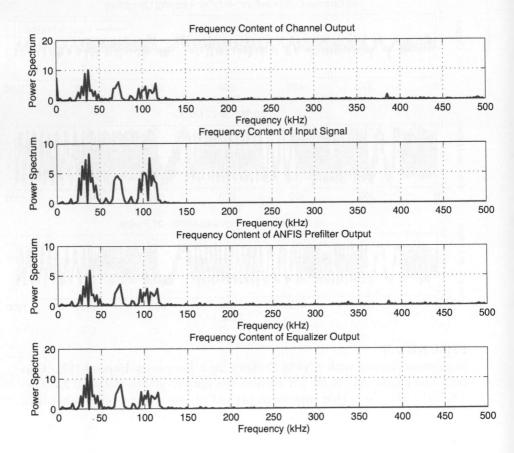

FIGURE 7.6
Spectra of Signals in Figure 7.5.

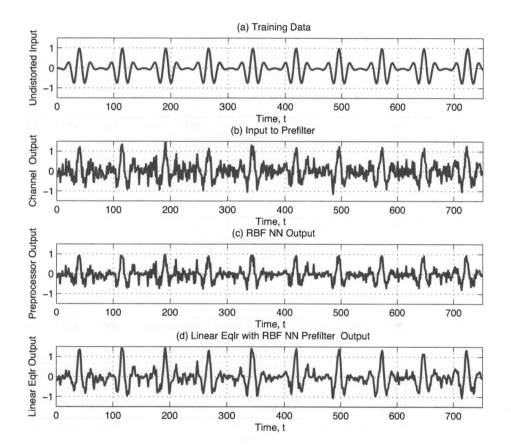

FIGURE 7.7

The signal $x(t)$ is a Gaussian pulse train; the output is given by $y(t) = x(t) + 0.2x^2(t) - 0.1x^3(t) + \eta(t)$. Note that waveforms (a) and (d) are closer to each other than (a) and (c).

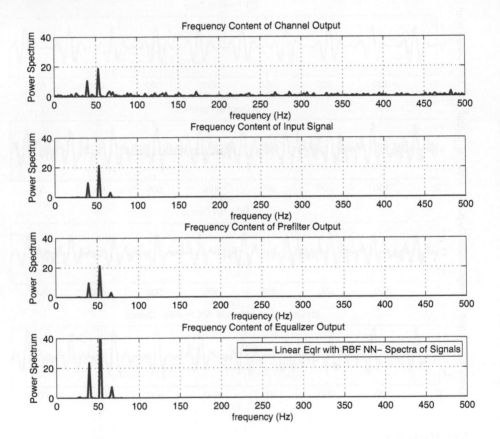

FIGURE 7.8

Spectra of Signals in Figure 7.7.

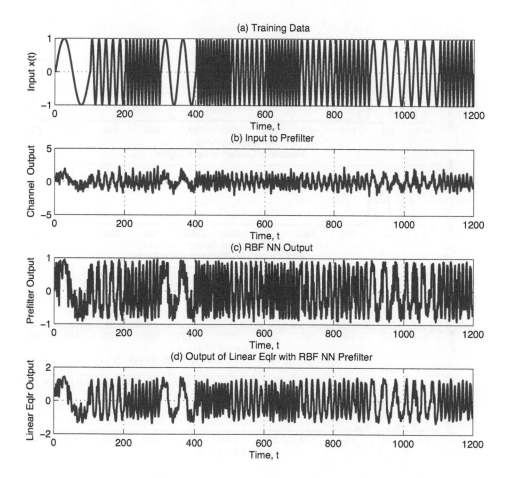

FIGURE 7.9
Simulation Results: RBF NN Prefilter with FH Carrier. Note that waveforms
(a) and (d) are closer to each other than (a) and (c).

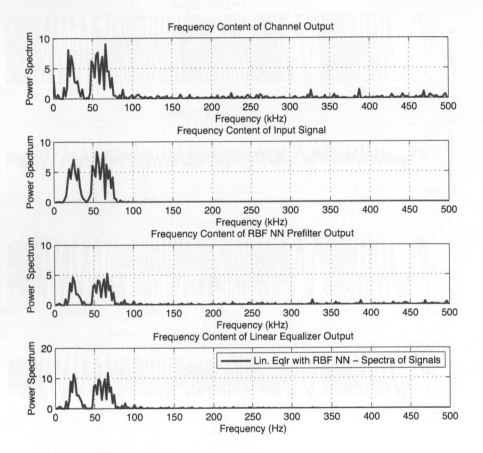

FIGURE 7.10
Spectra of Signals in Figure 7.9.

Theodore S. Rappaport, *Wireless Communications Principles and Practice*, Pearson Education, New Jersey 2003.

Hashemi H., The Indoor Radio Propagation Channel, *Proceedings of the IEEE*, Vol.81, No.7, pp.943–968, July 1993.

S.Y. Seidel and T.S. Rappaport, 914MHz Path Loss Prediction Models for Indoor Wireless Communications in Multifloored Buildings, *IEEE Transactions on Antennas and Propagation*, Vol.40, No.2, pp.207–217, February 1992.

B. Mulgrew, Applying Radial Basis Functions, *IEEE Signal Processing Magazine*, Vol.13, pp.50–65, March 1996.

A.P. Clark, I.H. Lee, and R.S. Marshall, Developments of the Conventional Non-linear Equaliser, *IEEE Proceedings*, Part F, Vol.129, No.2, pp.85–94, 1982.

D. Williamson, R.A. Kennedy, and G.W. Pulford, Block Decision Feedback Equalization, *IEEE Transactions on Communications*, Vol.40, No.2, pp.255–264, February 1992.

D.M.J Devasirvatham, A Comparison of Time-Delay Spread and Signal Level Measurements Within Two Dissimilar office Buildings, *IEEE Transactions on Antennas and Propagation*, Vol. AP-35, pp. 319–324, March 1987.

4

ANFIS-Based Channel Equalizer

The analysis of noise characteristics and modeling of a suitable equalizer for the nonlinear time-invariant mobile cellular channel is the focal theme of this book. This chapter is devoted to analyzing the functioning of the channel equalizer based on an Adaptive Network-based Fuzzy Inference System (ANFIS). It may be noted that the equalization of wireless mobile channels is a nonlinear problem, so a nonlinear solution is more appropriate.

We have to design the fuzzy *if-then else* rules based on the channel characteristics, namely, variances of signal, noise, co-channel (CCI) and adjacent channel interferences (ACI) as well as the transmitted signal (input)-received signal (output) mapping. The equalizer is a nonlinear system that effectively undoes the aberrations done to the transmitted signal by the channel due to the noise and co-channel and adjacent channel interferences. Now, modeling a nonlinear system is fairly complex so that conventional methods of system identification cannot be applied to find the inverse system. One possible experimental method to develop a model for indoor wireless channel (viz., the channel impulse response, CIR) is to carry out expensive channel sounding (for example, one could use the RUSK Channel sounder from RF Sub Systems, GmBH, which would cost over $100,000). In this book, we attempt to supplant the expensive channel sounding technique for mobile wireless channels (not restricted to the indoor case) by simulation.

The rest of the chapter is organized as follows. In Section 4.1, we introduce the working principle of ANFIS (Jang 1993).The methods of channel equalizer analysis and design are reviewed in Section 4.2. The mobile cellular channel equalizer based on ANFIS is introduced in Section 4.3, where we consider a number of equalizers based on ANFIS, with varying parameters. Thereafter, in Section 4.4, we consider the concept of Ultra-Wide Band (UWB) systems and their equalization using ANFIS. Conclusions are made in Section 4.5.

4.1 Introduction

System modeling techniques based on conventional mathematical tools like differential equations are not well suited for dealing with ill-defined and uncertain systems (Jang 1993). In contrast, a *fuzzy inference system (FIS)*, employing fuzzy if–then rules can model the qualitative aspects of human knowledge and reasoning processes without employing precise quantitative analyses. This *fuzzy modeling* or *fuzzy identification*, first explored systematically by Takagi and Sugeno (Jang 1993), has found numerous practical applications in control, prediction, and inference. However, there are some basic aspects of this approach which are in need of better understanding. More specifically, no standard methods exist for optimally transforming human knowledge or experience into the rule base and database of a fuzzy inference system. There is a need for effective methods for tuning the membership functions (MFs) so as to minimize the output error measure or maximize the performance index. In this perspective, a novel architecture called ANFIS, which can serve as a basis for constructing a set of fuzzy if–then rules with appropriate membership functions to generate the stipulated input–output pairs, is taken up (Jang 1993).

4.2   Methods of Channel Equalizer Analysis and Design

Adaptive filtering has achieved widespread application and success such as control, image processing, and communication (Williamson 1992, Widrow and Stearns 1991). Among the various adaptive filters, the adaptive linear filter is the most widely used mainly due to its low hardware implementation cost and other properties, like convergence, global minimum, misadjustment error, and training algorithms. It can be analyzed and derived easily. Adaptive linear filtering has achieved a large amount of success in many situations. The maximum likelihood sequence estimators (MLSE) (Forney 1972) are implemented using the Viterbi algorithm. The large computational complexity associated with the Viterbi algorithm and the poor performance of the linear equalizers have led to the development of symbol-by-symbol equalizers using the maximum a posteriori probability (MAP) principle – Bayesian equalizers (Lin and Ho 2004). These Bayesian equalizers have been approximated using nonlinear signal processing techniques like artificial neural networks (ANN) (Gibson et al. 1991, AlMashouq and Reed 1994), radial basis functions (RBF) (Chen et al. 1991, Chen et al. 1993), recurrent neural networks (Kechriotis and Manolakos 1994), and fuzzy filters (Liang and Mendel 2000, Wang and Mendel 1993, Lin and Juang 1994, Patra and Mulgrew 2000). The study of these new techniques can provide adaptive equalizers which have the advantages of both good performance and low computational cost (Lin and Juang 1994). Fuzzy filters are nonlinear filters that incorporate linguistic information in the form of IF–THEN fuzzy rules. Fuzzy filters have been used for equalization due to their success in the related area of pattern classification (Liang and Mendel 2000, Wang and Mendel 1993, Patra and Mulgrew 2000). Wang and Mendel (1993) present Fuzzy Basis Functions (FBF) for channel equalization. Lin and Juang (1994) have developed adaptive neuro fuzzy filters (ANFF) and use them for equalization and noise reduction. This ANFF constructs its rule base in a dynamic way with the training samples. Patra and Mulgrew (2000) have derived the close relationship between the fuzzy equalizers and the equalizer based on MAP. Liang and Mendel (2000) developed type-2 Fuzzy Adaptive Filters (FAF) and demonstrated that they could implement the Bayesian equalizer. The structures and learning algorithms of these models are both complicated and not suitable for practical implementation.

4.2.0.1 FIS

Fuzzy if–then rules or fuzzy conditional statements are expressions of the form *IF A THEN B*, where A and B are labels of fuzzy sets, characterized by appropriate membership functions (Jang 1993). Due to their concise form, fuzzy if–then rules are often employed to capture the imprecise modes of reasoning that play an essential role in the human ability to make decisions in an environment of uncertainty and imprecision. An example that describes a simple fact is

If pressure is high, then volume is small

where pressure and volume are linguistic variables, and *high* and *small* are *linguistic values* or *labels* that are characterized by membership functions (Zadeh 1973).

Another form of fuzzy if–then rule, proposed by Takagi and Sugeno, has fuzzy sets involved only in the premise part. By using Takagi and Sugeno's fuzzy if–then rule, we can describe the resistive force on a moving object as follows:

If velocity is high then $force = k \times velocity^2$,

where, again, *high* in the premise part is a linguistic label characterized by an appropriate membership function. However, the consequent part is described by a nonfuzzy equation of the input variable, *velocity*. Both types of fuzzy if–then rules have been used extensively in modeling and control. Basically a fuzzy inference system is composed of five functional blocks, as shown in Figure 4.1, which are

1. *a rule base* containing a number of fuzzy if–then rules.

2. *a database* which defines the membership functions of the fuzzy sets used in the fuzzy rules.

3. *a decision making unit* which performs the inference operations on the rules.

4. *a fuzzification interface* which transforms the crisp inputs into degrees of match with linguistic values.

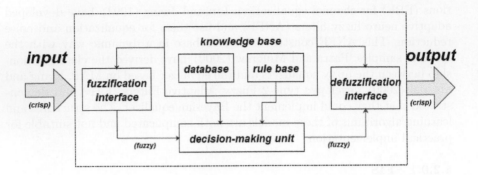

FIGURE 4.1
Block Diagram of an FIS. Note that the *Fuzzy Knowledge Base* Is Composed of a *Database* and, *Fuzzy Rule Base*.

5. *a defuzzification interface* which transform the fuzzy results of the inference into a crisp output.

Usually, the rule base and the database are jointly referred to as the *knowledge base*. The steps of *fuzzy reasoning* (inference operations upon fuzzy if–then rules) performed by fuzzy inference systems are:

1. Compare the input variables with the membership functions on the premise part to obtain the membership values for compatibility measures of each *linguistic label*. This step is often called fuzzification (Jang 1993).

2. Combine (through a specific *t-norm operator*, usually multiplication or min of the membership values on the premise part) to get the *firing strength (weight)* of each rule.

3. Generate the qualified consequent (either fuzzy or crisp) of each rule depending on the firing strength.

4. Aggregate the qualified consequents to produce a crisp output. This step is called *defuzzification* (Jang 1993).

Several types of fuzzy reasoning have been proposed in the literature. Depending on the types of fuzzy reasoning and fuzzy if–then rules employed, most fuzzy inference systems can be classified into three types (Jang 1993):

1. *Type-1*: The overall output is the weighted average of each rule's crisp output induced by the rule's firing strength (the product or minimum of the degrees of match with the premise part) and output membership functions. The output membership functions used in this scheme must be monotonically nondecreasing.

2. *Type-2*: The overall fuzzy output is derived by applying the *max* operation to the qualified fuzzy outputs, each of which is equal to the minimum of firing strength and the output membership function of each rule. Various schemes have been proposed to choose the final crisp output based on the overall fuzzy output; some of them are the *center of area*, the *bisector of area*, the *mean of maxima*, the *maximum criterion*, etc.

3. *Type-3*: Takagi and Sugeno's fuzzy if–then rules (TSK model) are used. The output of each rule is a linear combination of input variables plus a constant term, and the final output is the weighted average of each rule's output.

Commonly used fuzzy learning rules are illustrated in Figure 4.2.

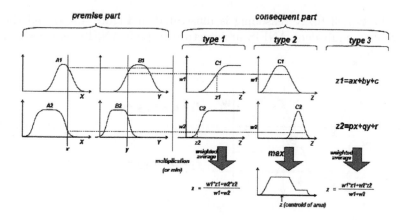

FIGURE 4.2
Various Fuzzy Reasoning Mechanisms: Type-1, Type-2, and Type-3. The difference is on the *consequent part*.

4.2.0.2 ANFIS

Functionally, there are almost no constraints on the node functions of an adaptive network except piecewise differentiability. Structurally, the only limitation on network configuration is that it should be of the feedforward type. Due to these minimal restrictions, the adaptive network's applications are immediate and immense in various areas. In this section, a class of adaptive networks which are functionally equivalent to fuzzy inference systems referred to as ANFIS, are examined.

Fuzzy modeling is applied to situations where the exact mathematical model is difficult to conceive and a measurement of values associated with the variables involved is quite tedious. Even the variables we considered in the present problem (i.e., variances of signal, noise, and co-channel and adjacent channel interferences) are themselves not measurable accurately. In such situations, fuzzy models are developed and used for precise estimation of the transmitted signal at the receiver side.

4.2.1   ANFIS Architecture and Functional Layers

For simplicity, we assume the fuzzy inference system under consideration has two inputs x and y and one output z (Jang 1993). Suppose that the rule base contains two fuzzy if–then rules of Takagi and Sugeno's type:

$$Rule\ 1:\quad If\ x\ is\ A_1\ and\ y\ is\ B_1,\quad then\ f_1 = p_1 x + q_1 y + r_1. \tag{4.1}$$

$$Rule\ 2:\quad If\ x\ is\ A_2\ and\ y\ is\ B_2,\quad then\ f_2 = p_2 x + q_2 y + r_2. \tag{4.2}$$

The the type-3 fuzzy reasoning is illustrated in Figure 4.3(a) and the corresponding equivalent ANFIS architecture (type-3 ANFIS) is shown in Figure 4.3(b).

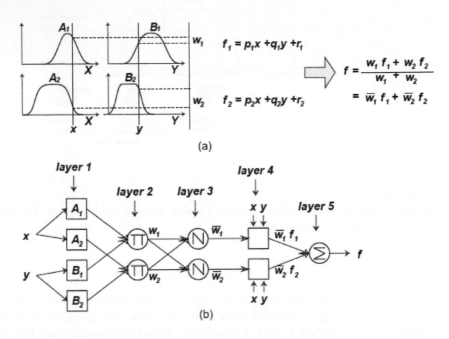

FIGURE 4.3
TSK Model of Fuzzy Inference System: (a) Type-3 Fuzzy Reasoning; (b) Equivalent ANFIS (Type-3 ANFIS).

4.2.1.1 Node Functions

The node functions in the same layer are of the same function family as described below (Jang 1993):

1. **Layer 1**: Every node i in this layer is a square node with a node function

$$O_i^1 = \mu_{A_i}(x), \tag{4.3}$$

where x is the input to node i, and A_i is the linguistic label (*small, large,* etc.) associated with this node function. In other words, O_i^1 is the membership function of A_i and it specifies the degree to which the given x satisfies the quantifier A_i. Usually we choose $\mu_{A_i}(x)$ to be bell-shaped with maximum equal to 1 and minimum equal to 0, such as

$$\mu_{A_i}(x) = \frac{1}{1 + \left[\left(\frac{x-c_i}{a_i}\right)^2\right]^{b_i}}, \tag{4.4}$$

or

$$\mu_{A_i}(x) = \exp\left\{-\left[\left(\frac{x-c_i}{a_i}\right)^2\right]^{b_i}\right\}, \tag{4.5}$$

where $\{a_i, b_i, c_i\}$ forms the parameter set. As the values of these parameters change, the bell-shaped functions vary accordingly, thus exhibiting various forms of membership functions on linguistic label A_i. In fact, any continuous and piecewise differentiable functions, such as commonly used trapezoidal or triangular-shaped membership functions, are also qualified candidates for node functions in this layer (Jang 1993). Parameters in this layer are referred to as *premise (or antecedent)* parameters.

2. **Layer 2**: Every node in this layer is a circle node labeled Π which multiplies the incoming signals and sends the product out (Jang 1993). For example,

$$w_i = \mu_{A_i}(x) \times \mu_{B_i}(y), \quad i = 1, 2. \tag{4.6}$$

Each node output represents the firing strength of a rule. In fact, other *t-norm* operators those perform generalized AND, can also be used as the node function in this layer.

3. **Layer 3**: Every node in this layer is a circle node labeled N. The ith node calculates the ratio of the ith rule's firing strength to the sum of all rules' firing strengths:

$$\overline{w}_i = \frac{w_i}{w_1 + w_2}, \quad i = 1, 2. \tag{4.7}$$

For convenience, outputs of this layer will be called *normalized firing strengths*.

4. **Layer 4**: Every node i in this layer is a square node with a node function

$$O_i^4 = \overline{w}_i \, f_i = \overline{w}_i (p_i x + q_i y + r_i), \tag{4.8}$$

where $\overline{w}_i$ is the output of layer 3, and $\{p_i, q_i, r_i\}$ is the parameter set. Parameters in this layer will be referred to as *consequent* parameters (Jang 1993).

5. **Layer 5**: The single node in this layer is a circle node labeled $\sum$ that computes the overall output as the summation of all incoming signals, i.e.,

$$O_i^5 = overall \ \ output = \sum_i \overline{w}_i \, f_i = \frac{\sum_i w_i f_i}{\sum_i w_i}. \tag{4.9}$$

4.3 Mobile Channel Equalizer Based on ANFIS

Since the equalization of a mobile cellular channel, which is basically a *nonlinear time-variant system*, is a nonlinear problem, a solution using ANFIS is most suitable for it. Here again, we have to choose a channel model as in Chapter 3, where we considered a FAF for channel equalization. For the ANFIS based equalizer, we use a type-3 TSK FIS with Gaussian membership functions. For a practical case, we choose five/seven rules for the input variables, each with a Gaussian membership function given by

$$\mu_{A_i}(x) = \exp \left\{ -\left(\frac{x - c_i}{a_i} \right)^{2b_i} \right\} \tag{4.10}$$

where $\{a_i, b_i, c_i\}$ is the parameter set. For a channel with 6 co-channels, (i.e., $N = 7$), we can consider the ANFIS equalizer as having 7 components in its input (plus the AWGN in the channel) and one output, which is connected to the ANFIS equalizer and detector, as shown in Figure 4.4.

4.3.1 Simulation of a Channel Equalizer Using MATLAB®

It has been found that a wireless channel can be modeled as nonlinear time-variant (NLTV) when the duration of the observation window is fairly long or as nonlinear time-invariant (NLTI) when the duration of the observation window is short. This fact is established by simulation, as it is a hard problem to obtain a rigorous mathematical proof.

Conventional channel models available in recent literature were studied to arrive at a suitable paradigm for the wireless channel, consisting of the different variables and parameters. This also enabled us to understand the inadequacies of existing mathematical models for wireless channels.

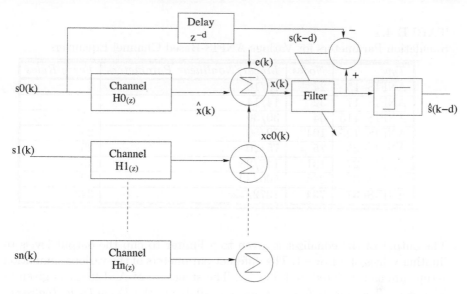

FIGURE 4.4
Discrete-Time Model of a Digital Communication System with AWGN and CCI.

The fuzzy if–then rules are generated by ANFIS based inverse system (to the channel), which effectively acts as an adaptive equalizer at the receiver side. The ANFIS automatically generates the rule base from a set of input-output data vectors. This is achieved by minimizing the error between actual input signal (at the transmitter of the wireless system) and the estimate of the input (at the receiver).

In the simulation, we assume that the external input to the ANFIS equalizer is the output of the channel, which is the sum of the desired channel output plus the weighted sum of the co-channel outputs and the Gaussian noise, which is assumed to be AWGN, with zero mean and standard deviation up to 0.8. In the ensuing sections, we use the following definitions for Signal-to-Noise Ratio (SNR), Signal-to-Interference Ratio (SIR), and Signal-to-Interference-Noise Ratio (SINR).

$$SNR = 10 \log_{10} \frac{\sigma_{\hat{s}}^2}{\sigma_{\hat{n}}^2} \qquad (4.11)$$

$$SIR = 10 \log_{10} \frac{\sigma_{\hat{s}}^2}{\sigma_{\hat{i}}^2} \qquad (4.12)$$

$$SINR = 10 \log_{10} \frac{\sigma_{\hat{s}}^2}{\sigma_{\hat{i}}^2 + \sigma_{\hat{n}}^2} \qquad (4.13)$$

where $\sigma_{\hat{s}}^2$, $\sigma_{\hat{n}}^2$, and $\sigma_{\hat{i}}^2$ are the variances of the signal, AWG noise, and the co-channel and adjacent channel interferences (put together) signal, respectively.

TABLE 4.1
Simulation Parameters for Various ANFIS-Based Channel Equalizers

Type	Nodes	Linear/Nonlinear Parameters	Fuzzy Rules
ANFIS–15	24	10/10	5
ANFIS–17	32	14/14	7
ANFIS–115	64	30/30	15
ANFIS–125	104	50/50	25
ANFIS–25	75	75/20	25
ANFIS–27	131	147/28	49
ANFIS–35	286	500/30	125
ANFIS–37	734	1372/42	343

The output of the equalizer is given to a limiter to clip the output levels to limiting values of $+1$ or -1. The different parameters of the various simulation setups are as tabulated in Table 4.1. The structure of ANFIS–27 is given in Figure 4.5. The library function, *anfis*, available in the *Fuzzy Logic Toolbox* of MATLAB® version 7.0 is used extensively in all simulations.

In Table 4.1 on simulation parameters for various ANFIS, the first digit in the ANFIS type (column 1) indicates the number of inputs to the ANFIS structure (as the **1** in ANFIS–115), and the following digit(s) indicate the number of fuzzy rules for each input(s). The last column indicates the total number of fuzzy rules for the entire ANFIS. The number of outputs is one in all cases.

Note that the ANFIS–27 based equalizer has two inputs from multipath components, seven fuzzy rules for each input, and one output that feeds the receiver subsystem.

4.3.2   Description of the ANFIS-Based Channel Equalizer

The Figure 4.6 shows the architecture of the proposed ANFIS based channel equalizer for seven fuzzy rules. The wireless channel modeling based on artificial neural networks is capable of depicting the input-output mapping existing in the equalizer system and it does provide us with an exact picture of the variables and parameters defining the system. Moreover, neural network based models do have learning capability.

Fuzzy models, on the other hand, do not possess learning capability. Therefore, fusing together these two, we can have a model which is capable of both depicting the dynamics of the system in terms of the variables and parameters has the self-learning capability. The adaptability of the equalizer under purview is achieved by the learning aspect of neural network. The fuzzy reasoning (especially the TSK model used in ANFIS) maps the input to the output. We follow a first-order ANFIS with the antecedent parameters being the standard deviations of the received signal, CCI and ACI interferences (put

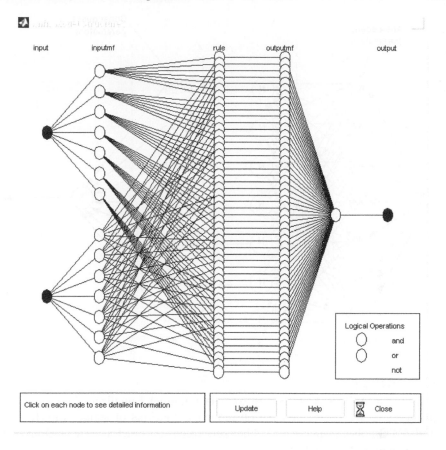

FIGURE 4.5
Structure of ANFIS–27 Generated Using MATLAB Fuzzy Logic Toolbox: Number of Inputs = 2, Number of Outputs = 1, Total Number of Fuzzy Rules = 49, Type of Membership Function: Gaussian, and Number of Nodes = 131.

together), and the AWGN ($\sigma_{\hat{s}}$, $\sigma_{\hat{i}}$, and $\sigma_{\hat{n}}$, respectively), collectively represented as A_i. The only consequent parameter is the scaling factor of the signal (ρ_i) at the output. The membership functions of A_i, $i = 1, 2, \ldots, 7$ are chosen to be *Gaussian*. Some of the rules in the *fuzzy rule base* can be stated as

$$If\ \sigma_{\hat{s}}\ is\ \textbf{very low},\ and\ \sigma_{\hat{i}}\ is\ \textbf{very low},$$
$$and\ \sigma_{\hat{n}}\ is\ \textbf{very low}\ then\ y = \rho_1 s. \quad (4.14)$$

$$If\ \sigma_{\hat{s}}\ is\ \textbf{low},\ and\ \sigma_{\hat{i}}\ is\ \textbf{very low},\ and\ \sigma_{\hat{n}}\ is\ \textbf{very low}\ then\ y = \rho_2 s. \quad (4.15)$$

$$If\ \sigma_{\hat{s}}\ is\ \textbf{medium},\ and\ \sigma_{\hat{i}}\ is\ \textbf{very low},$$
$$and\ \sigma_{\hat{n}}\ is\ \textbf{very low}\ then\ y = \rho_3 s. \quad (4.16)$$

$$If\ \sigma_{\hat{s}}\ is\ \textbf{medium},\ and\ \sigma_{\hat{i}}\ is\ \textbf{low},\ and\ \sigma_{\hat{n}}\ is\ \textbf{low}\ then\ y = \rho_4 s. \quad (4.17)$$

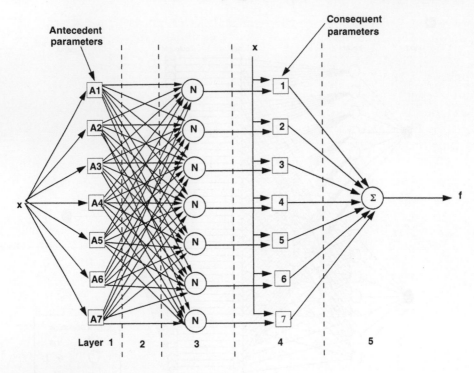

FIGURE 4.6
Equivalent ANFIS Architecture for the Channel Equalizer. The antecedent parameters are $\sigma_{\hat{s}}, \sigma_{\hat{i}}$, and $\sigma_{\hat{n}}$ (standard deviations of signal, interferences, and AWG noise, respectively). The consequent parameter is the scaling factor of the signal at the output, ρ.

The three input variables can assume any one of the 5 possible membership functions from the set {*very low, low, medium, high, very high*}, leaving us with 125 possible combinations of rules. However, using *fuzzy rule reduction techniques* the total number of rules can be limited to 7 or 25. The overall output of y is given by

$$y = \sum_{i=1}^{7} [\mu_{A_i}(s).\rho_i s] \, / \sum_{i=1}^{7} \mu_{A_i}(s). \tag{4.18}$$

The steps in the algorithm for simulation of the ANFIS–27 based equalizer are as given below.

1. The standard deviations of CCI and AWGN are logarithmically varied from 0.02 to 0.8. This information is derived from the literature.

2. The random binary input data (which represents the input to the channel

from the transmitter) is generated and the corrupted data available at the outputs of the two multipaths due to CCI and AWGN is obtained.

3. Set the number of membership functions as 7, membership function type as *"Gaussian"*, and the number of epochs to 80.

4. Simulate the ANFIS (which implements the equalizer) and plot the results.

The error plot of the ANFIS–27 training is illustrated in Figure 4.7. We have set the number of epochs as 80 in this case.The ANFIS-27 consists of 2 inputs and one output, and 7 fuzzy rules for each membership function. The fuzzy membership functions are chosen to be Gaussian.

4.3.3   Results of Simulations

The output of the channel (received signal), which is a nonlinear combination of the signal, the co-channel signals, and the AWG noise, is a random waveform taking values around $+1$ and -1, as seen from the simulated waveform, shown in Figure 4.8. The equalized, output after thresholding, will be very much identical as shown in Figure 4.8. The simulation results for ANFIS–23 (with 2 inputs and 3 membership functions), ANFIS–25 (with 2 inputs and 5 membership functions), and ANFIS–27 (with 2 inputs and 7 membership functions) for 4096 training data pairs are shown in Figure 4.8, as a plot of the *time domain response*.

The MATLAB code used for the simulation, illustrated in Figure 4.8, is appended below.

```
%%% ANFIS2x Eqlr Simulation with 2 inputs and
%%% 3/5/7 membership functions and 4096 data pairs.
%%% Last modified on 22-10-2012.
clear all;clf;close all;clc;
tic;
ns=1024;
nb=4;
t=[1:ns*nb];
[x,b] = random_binary(ns,nb);
[x1,b1] = random_binary(ns,nb);
[x2,b2] = random_binary(ns,nb);
[x3,b3] = random_binary(ns,nb);
[x4,b4] = random_binary(ns,nb);
[x5,b5] = random_binary(ns,nb);
[x6,b6] = random_binary(ns,nb);
e1=0.2*randn(ns*nb,1);
e2=0.2*randn(ns*nb,1);
y1 = x'+0.2*(x1'+x2'+x3'+x4'+x5'+x6')+e1;
y2 = x'+0.2*(x1'+x2'+x3'+x4'+x5'+x6')+e2;
```

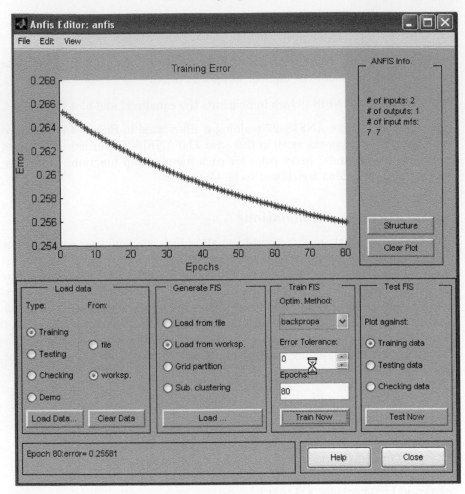

FIGURE 4.7 (See color insert.)
Error Plot of Training of ANFIS–27; Generated Using MATLAB Fuzzy Logic
Toolbox: Number of Inputs = 2, Number of Outputs = 1, Total Number of
Fuzzy Rules = 49, Type of Membership Function: Gaussian, and Number of
Epochs = 80.

```
y=[y1 y2];
trnData = [y x'];
numMFs = 3;
mfType = 'gaussmf';
epoch_n = 20;
in_fismat = genfis1(trnData,numMFs,mfType);
out_fismat = anfis(trnData,in_fismat,20);
est_x23=evalfis(y,out_fismat);
```

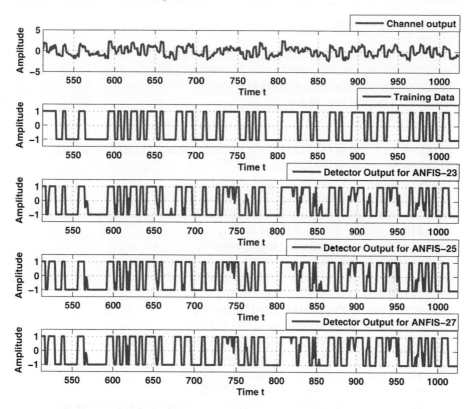

FIGURE 4.8 (See color insert.)
Simulation Results for ANFIS–23, ANFIS–25, and ANFIS–27 Equalizers for 4096 Training Data Pairs, showing the *Time Domain Response*. The plots at the bottom represent the output of ANFIS–23, ANFIS–25, and ANFIS–27 equalizers with an attached hard thresholding detector.

```
est_x23(est_x23<-0.6)=-1.0;
est_x23(est_x23>0.6)=1.0;
%%%%
numMFs = 5;
mfType = 'gaussmf';
epoch_n = 20;
in_fismat = genfis1(trnData,numMFs,mfType);
out_fismat = anfis(trnData,in_fismat,20);
est_x25=evalfis(y,out_fismat);
est_x25(est_x25<-0.6)=-1.0;
est_x25(est_x25>0.6)=1.0;
%%%%
numMFs = 7;
```

```
mfType = 'gaussmf';
epoch_n = 20;
in_fismat = genfis1(trnData,numMFs,mfType);
out_fismat = anfis(trnData,in_fismat,20);
est_x27=evalfis(y,out_fismat);
est_x27(est_x27<-0.6)=-1.0;
est_x27(est_x27>0.6)=1.0;
%%%
subplot(511),plot(t(512:1024),y1(512:1024),'LineWidth',2);
axis([512 1024 -5 5]);grid on;
xlabel('Time t');ylabel('Amplitude');
legend('Channel output');
subplot(512),plot(t(512:1024),x(512:1024),'LineWidth',2);
axis([512 1024 -1.5 1.5]);grid on;
xlabel('Time t');ylabel('Amplitude');
legend('Training Data');
subplot(513),plot(t(512:1024),est_x23(512:1024),'LineWidth',2);
axis([512 1024 -1.5 1.5]);grid on;
xlabel('Time t');ylabel('Amplitude');
legend('Detector Output for ANFIS-23');
subplot(514),plot(t(512:1024),est_x25(512:1024),'LineWidth',2);
axis([512 1024 -1.5 1.5]);grid on;
xlabel('Time t');ylabel('Amplitude');
legend('Detector Output for ANFIS-25');
subplot(515),plot(t(512:1024),est_x27(512:1024),'LineWidth',2);
axis([512 1024 -1.5 1.5]);grid on;
xlabel('Time t');ylabel('Amplitude');
legend('Detector Output for ANFIS-27');
toc;
%%% end of anfis2xEq.m
```

Results for other combinations of number of inputs and membership functions, as listed in Table 4.1, are found to be similar. The processing times in each case for 1024 training data pairs are tabulated in Table 4.2.[1] In one of the simulations, the standard deviation of CCI and AWGN is logarithmically varied between 0.02 and 0.8 using the MATLAB command ($[logspace(log10(0.02), log10(0.8), 16)]$) and the simulation is run on a total of 2048/4096 training data pairs. The results are shown in Figure 4.9, as a plot of $log(BER)$ at the output of the equalizer versus SINR in dB. Then, in another simulation, the $log(BER)$ at the output of the equalizer is calculated for standard deviation of noise varying from 0.02 to 0.8 for two versions of ANFIS equalizers (ANFIS–115 and ANFIS–125) for 2048/4096 training data pairs and standard deviation of AWGN fixed at 0.42, and the results are plot-

---

[1] The simulations were run on a personal computer with an Intel Pentium 4 CPU running at 2.6 GHz and 512 MB RAM, using MATLAB version 7.0 software.

ted in Figure 4.10. The performance for the above ANFIS pairs, as regards log(BER) at the output of the equalizer versus SNR in dB for standard deviation of co-channel interference signal fixed at 0.08 is given in Figure 4.11.

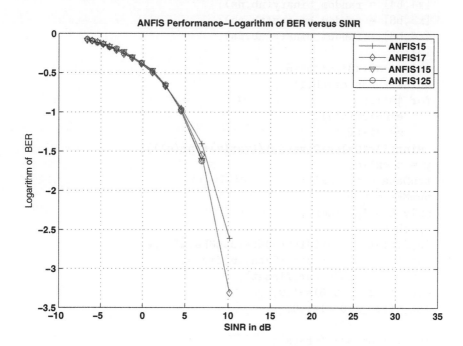

FIGURE 4.9 (See color insert.)
Performance of ANFIS Equalizers. Logarithm of BER at the output of the equalizer versus SINR in dB (varies from −26 to −8).

The MATLAB script for the above simulation is given below:

```
%%% Modified ANFIS Equalizer Simulation with more
%%% precision. Plots Logarithm of BER versus SINR
%%% std of CCI varied from 0.02 to 0.8.
%%% std of  AWGN varied from 0.02 to 0.8.
%%% last modified on 22-10-2012.
%anfis15.m
clc;clf;clear all;close all;
tic;
nb=1024;
ns=4;
it=16;
% t=linspace(0.02,0.8,it);
t=logspace(log10(0.02),log10(0.8),it);
[x,b] = random_binary(nb,ns);
```

```
[x1,b1] = random_binary(nb,ns);
[x2,b2] = random_binary(nb,ns);
[x3,b3] = random_binary(nb,ns);
[x4,b4] = random_binary(nb,ns);
[x5,b5] = random_binary(nb,ns);
[x6,b6] = random_binary(nb,ns);
i=1;
cci=x1+x2+x3+x4+x5+x6;
noise=randn(ns*nb,1);
for j=1:it
    e=noise*t(j);
    z=cci*t(j);
 sinr(i)=10*log10(var(x)/(var(z)+var(e)));
y = x'+z'+e;
trnData = [y    x'];
numMFs = 5;
mfType = 'gaussmf';
epoch_n = 20;
in_fismat = genfis1(trnData,numMFs,mfType);
out_fismat = anfis(trnData,in_fismat,20);
est_x=evalfis(y,out_fismat);
est_x(est_x<-0.6)=-1.0;
est_x(est_x>0.6)=1.0;
ec=(x~=est_x');
ber(i)=sum(ec)/(nb*ns);
i=i+1;
end;
plot(sinr, log10(ber),'-+');hold on;grid on;
xlabel('SINR in dB');
ylabel('Logarithm of BER');
%end of ANFIS15;
i=1;
for j=1:it
    e=noise*t(j);
    z=cci*t(j);
 sinr(i)=10*log10(var(x)/(var(z)+var(e)));
y = x'+z'+e;
trnData = [y    x'];
numMFs = 7;
mfType = 'gaussmf';
epoch_n = 20;
in_fismat = genfis1(trnData,numMFs,mfType);
out_fismat = anfis(trnData,in_fismat,20);
est_x=evalfis(y,out_fismat);
est_x(est_x<-0.6)=-1.0;
```

```
est_x(est_x>0.6)=1.0;
ec=(x~=est_x');
ber(i)=sum(ec)/(nb*ns);
i=i+1;
end;
plot(sinr, log10(ber),'-d');hold on;grid on;
xlabel('SINR in dB');
ylabel('Logarithm of BER');
%end of ANFIS17;
i=1;
for j=1:it
    e=noise*t(j);
    z=cci*t(j);
 sinr(i)=10*log10(var(x)/(var(z)+var(e)));
y = x'+z'+e;
trnData = [y    x'];
numMFs = 15;
mfType = 'gaussmf';
epoch_n = 20;
in_fismat = genfis1(trnData,numMFs,mfType);
out_fismat = anfis(trnData,in_fismat,20);
est_x=evalfis(y,out_fismat);
est_x(est_x<-0.6)=-1.0;
est_x(est_x>0.6)=1.0;
ec=(x~=est_x');
ber(i)=sum(ec)/(nb*ns);
i=i+1;
end;
plot(sinr, log10(ber),'-v');hold on;grid on;
xlabel('SINR in dB');
ylabel('Logarithm of BER');
%end of ANFIS115;
i=1;
for j=1:it
    e=noise*t(j);
    z=cci*t(j);
 sinr(i)=10*log10(var(x)/(var(z)+var(e)));
y = x'+z'+e;
trnData = [y    x'];
numMFs = 25;
mfType = 'gaussmf';
epoch_n = 20;
in_fismat = genfis1(trnData,numMFs,mfType);
out_fismat = anfis(trnData,in_fismat,20);
est_x=evalfis(y,out_fismat);
```

```
est_x(est_x<-0.6)=-1.0;
est_x(est_x>0.6)=1.0;
ec=(x~=est_x');
ber(i)=sum(ec)/(nb*ns);
i=i+1;
end;
plot(sinr, log10(ber),'-o');hold on;grid on;
xlabel('SINR in dB');
ylabel('Logarithm of  BER');
%end of ANFIS125;
legend('ANFIS15','ANFIS17','ANFIS115','ANFIS125');
title('ANFIS Performance-Logarithm of BER versus SINR');
hold off;
toc;
%%%end of anExEx2.m
```

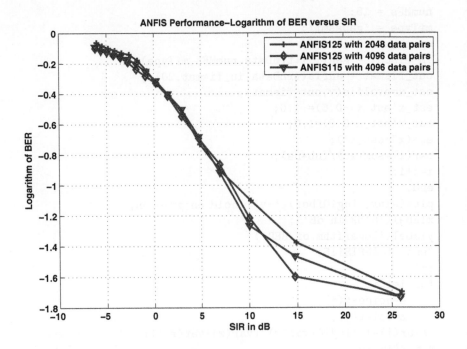

FIGURE 4.10 (See color insert.)
Performance of ANFIS Equalizers. Logarithm of BER at the output of the equalizer versus SIR in dB (varies from −10 to 30, with standard deviation of AWGN fixed at 0.42).

The MATLAB script for the simulation illustrated in Figure 4.10 is appended below.

```
%%% Modified ANFIS Equalizer Simulation with more
%%% precision. Plots Logarithm of BER versus SIR
%%% std of CCI varied from 0.02 to 0.8.
%%% std of AWGN fixed to 0.42.
%%% last modified on 22-10-2012..
%anfis125.m
clc;clf;clear all;close all;clf;
tic;
nb=512;
ns=4;
it=16;
t=linspace(0.02,0.8,it);
%t=logspace(log10(0.02),log10(0.8),it);
[x,b] = random_binary(nb,ns);
[x1,b1] = random_binary(nb,ns);
[x2,b2] = random_binary(nb,ns);
[x3,b3] = random_binary(nb,ns);
[x4,b4] = random_binary(nb,ns);
[x5,b5] = random_binary(nb,ns);
[x6,b6] = random_binary(nb,ns);
i=1;
cci=x1+x2+x3+x4+x5+x6;
e=randn(ns*nb,1)*0.42;
for j=1:it
    z=cci*t(j);
 sir(i)=10*log10(var(x)/var(z));
y = x'+z'+e;
trnData = [y    x'];
numMFs = 25;
mfType = 'gaussmf';
epoch_n = 20;
in_fismat = genfis1(trnData,numMFs,mfType);
out_fismat = anfis(trnData,in_fismat,20);
est_x=evalfis(y,out_fismat);
est_x(est_x<-0.6)=-1.0;
est_x(est_x>0.6)=1.0;
ec=(x~=est_x');
ber(i)=sum(ec)/(nb*ns);
i=i+1;
end;
plot(sir, log10(ber),'-+');hold on;grid on;
xlabel('SIR in dB');
ylabel('Logarithm of BER');
% end of ANFIS125 with 2048 data pairs..
%%% anfis125.m with 4096 data pairs..
```

```
nb=1024;
ns=4;
it=16;
t=linspace(0.02,0.8,it);
%t=logspace(log10(0.02),log10(0.8),it);
[x,b] = random_binary(nb,ns);
[x1,b1] = random_binary(nb,ns);
[x2,b2] = random_binary(nb,ns);
[x3,b3] = random_binary(nb,ns);
[x4,b4] = random_binary(nb,ns);
[x5,b5] = random_binary(nb,ns);
[x6,b6] = random_binary(nb,ns);
i=1;
cci=x1+x2+x3+x4+x5+x6;
e=randn(ns*nb,1)*0.42;
for j=1:it
    z=cci*t(j);
 sir(i)=10*log10(var(x)/var(z));
    y = x'+z'+e;
trnData = [y    x'];
numMFs = 25;
mfType = 'gaussmf';
epoch_n = 20;
in_fismat = genfis1(trnData,numMFs,mfType);
out_fismat = anfis(trnData,in_fismat,20);
est_x=evalfis(y,out_fismat);
est_x(est_x<-0.6)=-1.0;
est_x(est_x>0.6)=1.0;
ec=(x~=est_x');
ber(i)=sum(ec)/(ns*nb);
i=i+1;
end;
plot(sir, log10(ber),'-d');hold on; grid on;
%end of ANFIS125.m
%% anfis115.m with 4096 data pairs..
i=1;
cci=x1+x2+x3+x4+x5+x6;
e=randn(ns*nb,1)*0.42;
for j=1:it
    z=cci*t(j);
 sir(i)=10*log10(var(x)/var(z));
    y = x'+z'+e;
trnData = [y x'];
numMFs = 15;%number of membership_rules
mfType = 'gaussmf';
```

```
epoch_n = 20;
in_fismat = genfis1(trnData,numMFs,mfType);
out_fismat = anfis(trnData,in_fismat,20);
est_x=evalfis(y,out_fismat);
est_x(est_x<-0.6)=-1.0;
est_x(est_x>0.6)=1.0;
ec=(x~=est_x');
ber(i)=sum(ec)/(nb*ns);
i=i+1;
end;
plot(sir,log10(ber),'-v');hold on;grid on;
%end of ANFIS115.m
ch1=['ANFIS125 with 2048 data pairs'];
ch2=['ANFIS125 with 4096 data pairs'];
ch3=['ANFIS115 with 4096 data pairs'];
legend(ch1,ch2,ch3);
title('ANFIS Performance-Logarithm of BER versus SIR');
hold off;
toc;
%%%end of anEnEx.m
```

The MATLAB script for the simulation illustrated in Figure 4.11 is given below.

```
%%% Modified ANFIS Equalizer Simulation for diff.
%%% number of data pairs with more precision.
%%% Plots BER versus SNR
%%% std of CCI fixed at 0.18.
%%% std of  AWGN varied from 0.02 to 0.8.
%%% Last modified on 22-10-2012..
%ANFIS125.m
clc;clf;clear all;close all;
tic;
nb=512;
ns=4;
it=16;
t=logspace(log10(0.02),log10(0.8),it);%%
[x,b] = random_binary(nb,ns);
[x1,b1] = random_binary(nb,ns);
[x2,b2] = random_binary(nb,ns);
[x3,b3] = random_binary(nb,ns);
[x4,b4] = random_binary(nb,ns);
[x5,b5] = random_binary(nb,ns);
[x6,b6] = random_binary(nb,ns);
i=1;
cci=x1+x2+x3+x4+x5+x6;
```

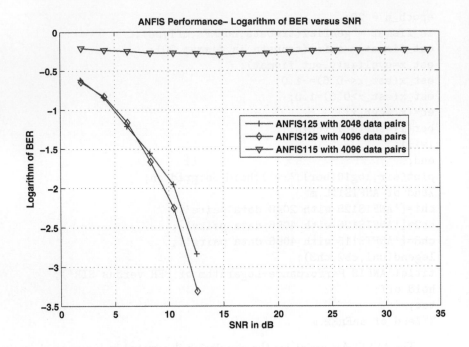

FIGURE 4.11 (See color insert.)
Simulation Results: Logarithm of BER at the output of the equalizer (varies
from 0 to −3.5) versus SNR in dBs (varies from 0 to 35) for 2048/4096 training
data pairs with standard deviation of CCI fixed at 0.08.

```
noise=randn(ns*nb,1);
for j=1:it
    e=noise*t(j);
    z=cci*0.08; % std of cci is fixed as 0.08
  snr(i)=10*log10(var(x)/var(e));
y = x'+z'+e;
trnData = [y    x'];
numMFs = 25;
mfType = 'gaussmf';
epoch_n = 20;
in_fismat = genfis1(trnData,numMFs,mfType);
out_fismat = anfis(trnData,in_fismat,20);
est_x=evalfis(y,out_fismat);
est_x(est_x<-0.6)=-1.0;
est_x(est_x>0.6)=1.0;
ec=(x~=est_x');
ber(i)=sum(ec)/(nb*ns);
```

```
i=i+1;
end;
plot(snr, log10(ber),'-+');hold on;grid on;
xlabel('SNR in dB');
ylabel('Logarithm of BER');
%end of ANFIS125;
nb=1024;
ns=4;
it=16;
t=logspace(log10(0.02),log10(0.8),it);
[x,b] = random_binary(nb,ns);
[x1,b1] = random_binary(nb,ns);
[x2,b2] = random_binary(nb,ns);
[x3,b3] = random_binary(nb,ns);
[x4,b4] = random_binary(nb,ns);
[x5,b5] = random_binary(nb,ns);
[x6,b6] = random_binary(nb,ns);
i=1;
cci=x1+x2+x3+x4+x5+x6;
noise=randn(ns*nb,1);
for j=1:it
    e=noise*t(j);
    z=cci*0.08; % std of cci is fixed as 0.18
 snr(i)=10*log10(var(x)/var(e));
y = x'+z'+e;
trnData = [y    x'];
numMFs = 25;
mfType = 'gaussmf';
epoch_n = 20;
in_fismat = genfis1(trnData,numMFs,mfType);
out_fismat = anfis(trnData,in_fismat,20);
est_x=evalfis(y,out_fismat);
est_x(est_x<-0.6)=-1.0;
est_x(est_x>0.6)=1.0;
ec=(x~=est_x');
ber(i)=sum(ec)/(nb*ns);
i=i+1;
end;
plot(snr, log10(ber),'-d');hold on;grid on;
xlabel('SNR in dB');
ylabel('Logarithm of BER');
%end of ANFIS125;
%%anfis115.m
it=16;
t=logspace(log10(0.02),log10(0.8),it);
```

```
[x,b] = random_binary(nb,ns);
[x1,b1] = random_binary(nb,ns);
[x2,b2] = random_binary(nb,ns);
[x3,b3] = random_binary(nb,ns);
[x4,b4] = random_binary(nb,ns);
[x5,b5] = random_binary(nb,ns);
[x6,b6] = random_binary(nb,ns);
i=1;
for j=1:it
    e=noise*t(j);
    z=cci*.08;
 snr(i)=10*log10(var(x)/var(e));
    y1 = x'+z'+e;
trnData = [y1 x'];
numMFs = 15;%number of membership_rules
mfType = 'gaussmf';
epoch_n = 20;
in_fismat = genfis1(trnData,numMFs,mfType);
out_fismat = anfis(trnData,in_fismat,20);
est_x=evalfis(y,out_fismat);
est_x(est_x<-0.6)=-1.0;
est_x(est_x>0.6)=1.0;
ec=(x~=est_x');
ber(i)=sum(ec)/(nb*ns);
i=i+1;
end;
plot(snr,log10(ber),'-v');hold on;grid on;
%end of ANFIS115.m
ch1=['ANFIS125 with 2048 data pairs'];
ch2=['ANFIS125 with 4096 data pairs'];
ch3=['ANFIS115 with 4096 data pairs'];
legend(ch1,ch2,ch3);
title('ANFIS Performance- Logarithm of BER versus SNR');
hold off;
toc;
%%% end of anEX1k4k.m
```

A plot of the performance of two different ANFIS structures (average BER at the output of the equalizer versus SNR and standard deviation of BER at the output of the equalizer versus standard deviation of AWGN) based on 100 Monte-Carlo (MC) simulations is given in Figure 4.12 for ANFIS–115 and ANFIS–125 structures; 1024 training data pairs are used in the simulation.

The MATLAB script file used for the simulation illustrated in Figure 4.12 is appended below

```
%%% Program to plot the Mean and std of BER
```

TABLE 4.2

Simulation Time for ANFIS with 1024 Training Data Pairs

ANFIS Type	Number of Epochs	Time in Seconds
ANFIS–15	20	0.84
ANFIS–17	20	1.04
ANFIS–115	20	2.34
ANFIS–125	20	5.50
ANFIS–25	20	7.85
ANFIS–27	20	26.80
ANFIS–35	20	558.80
ANFIS–37	20	4255.64

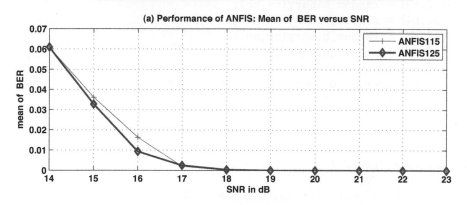

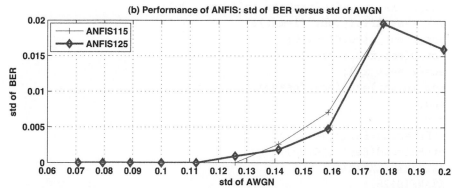

FIGURE 4.12 (See color insert.)

Performance of ANFIS Equalizers: (a) Mean BER at the output of the equalizer (varies from 0 to 0.07) versus SNR in dBs (varies from 14 to 23), and (b) standard deviation of BER at the output of the equalizer (varies from 0 to 0.02) versus standard deviation of AWGN (varies from 0.06 to 0.2).

```
%%% versus    SNR  and std of AWGN, respectively
%%% after 10 MC simulation. Last modified on 22-10-2012.
% anfis115.m
clc; clear all;close all;clf;
tic;
ns=256;% number of symbols.
nb=4;% number of bits per symbol
for mc=1:10% for 10 MC simulations.
for snr=14:23%for 10 values of SNR.
[x,b] = random_binary(ns,nb);
[x1,b1] = random_binary(ns,nb);
[x2,b2] = random_binary(ns,nb);
[x3,b3] = random_binary(ns,nb);
[x4,b4] = random_binary(ns,nb);
[x5,b5] = random_binary(ns,nb);
[x6,b6] = random_binary(ns,nb);
z=x1+x2+x3+x4+x5+x6;
s=sqrt(var(x)/(10^(snr/10))); %std of AWGN
e=s*randn(ns*nb,1);
y = x'+s*z'+e;
trnData = [y    x'];
numMFs = 15;
mfType = 'gaussmf';
epoch_n = 20;
in_fismat = genfis1(trnData,numMFs,mfType);
out_fismat = anfis(trnData,in_fismat,20);
est_x=evalfis(y,out_fismat);
est_x(est_x<-0.6)=-1.0;
est_x(est_x>0.6)=1.0;
ec=(x~=est_x');
Ber(mc,snr)=sum(ec)/(ns*nb);
end;
end;
mBER1=mean(Ber(:,14:23));
stBER1=std(Ber(:,14:23));
%%%%%%%%%%%%%%%
%%ANFIS125
for mc=1:10% for 10 MC simulations.
for snr=15:23%for 9 values of SNR.
[x,b] = random_binary(ns,nb);
[x1,b1] = random_binary(ns,nb);
[x2,b2] = random_binary(ns,nb);
[x3,b3] = random_binary(ns,nb);
[x4,b4] = random_binary(ns,nb);
[x5,b5] = random_binary(ns,nb);
```

```
[x6,b6] = random_binary(ns,nb);
z=x1+x2+x3+x4+x5+x6;
s=sqrt(var(x)/(10^(snr/10))); %std of AWGN
e=s*randn(ns*nb,1);
y = x'+s*z'+e;
trnData = [y    x'];
numMFs = 25;
mfType = 'gaussmf';
epoch_n = 20;
in_fismat = genfis1(trnData,numMFs,mfType);
out_fismat = anfis(trnData,in_fismat,20);
est_x=evalfis(y,out_fismat);
est_x(est_x<-0.6)=-1.0;
est_x(est_x>0.6)=1.0;
ec=(x~=est_x');
Ber(mc,snr)=sum(ec)/(ns*nb);
end;
end;
mBER2=mean(Ber(:,14:23));
stBER2=std(Ber(:,14:23));
snr=[14:23];
subplot(211), plot(snr,mBER1,'-+');hold on
plot(snr,mBER2,'-d'); grid on;
xlabel('SNR in dB');
ylabel('mean of  BER');
legend('ANFIS115','ANFIS125');
title('Performance of ANFIS: Mean of  BER versus SNR');
hold off;
st=sqrt(var(x)./(10.^(snr/10)));
subplot(212),plot(st,stBER1,'-+'); hold on;
plot(st,stBER2,'-d'); grid on;
xlabel('std of AWGN');
ylabel('std of  BER');
legend('ANFIS115','ANFIS125');
ch1=['Performance of ANFIS:'];
ch2=[' std of  BER versus std of AWGN'];
ch=strcat(ch1,ch2);
title(ch);
hold off;
toc;
%%% end of anfisexp.m
```

4.3.4 Interpretation of Results and Observations

The following observations are made based on Figures 4.9, 4.10, 4.11, and 4.12 and Tables 4.1 and 4.2 as well as results of simulations with fewer of data pairs.

1. With a larger number of training data pairs, BER at the output of the equalizer is reduced. This is due to the fact that the ANFIS gets optimally tuned with more training data pairs.

2. As the number of rules are increased, the BER at the output of the equalizer is reduced. A finer control is effected by increasing the number of rules, thereby reducing the BER. But this is attained at the cost of more time for ANFIS training.

3. As shown in Figure 4.9, performance of all ANFIS equalizers w.r.t. $\log(BER)$ at the output of the equalizer versus SINR, is nearly identical. When the SINR is above −10 dB, practically the $\log(BER)$ becomes close to zero. However, ANFIS–125 performs slightly better than other structures.

4. The performance of ANFIS–125 w.r.t. $\log(BER)$ at the output of the equalizer versus SIR is almost identical with 2048 or 4096 data pairs. However, for ANFIS–115, performance is slightly worse.

5. As shown in Figure 4.11, the performance of ANFIS–125 w.r.t. $\log(BER)$ at the output of the equalizer versus SNR is almost identical with 2048 or 4096 data pairs. However, for ANFIS–115, performance is very poor even at an SNR of 35 dB.

6. As we increase the number of rules or the number of inputs applied in parallel to the ANFIS structure, the number of internal nodes and the ANFIS training time increase. This is because, with more rules or more internal inputs to the ANFIS, the system can be modeled more accurately.

7. For MISO or MIMO systems, increasing the number of membership functions is the option for accurate system modeling, since in these cases the number of inputs applied to the ANFIS is two or more, and hence it will not be optimal to increase the number of internal inputs in the ANFIS.

8. An optimal ANFIS structure can be obtained based on the training time and the maximum error that can be tolerated. As indicated in Figure 4.12, at higher values of standard deviation of AWGN, and that of standard deviation of BER will be less with more membership functions. Hence the standard deviation of BER at the output of the equalizer can be yet another criterion in selecting a particular ANFIS structure.

Recently, a *modified ANFIS* (MANFIS) structure was presented by Jovanovic et al., which was shown to be more efficient with regard to root mean square error (RMSE) and computation time (Jovanovic et al. 2004), with a mutual trade-off. This essentially improves the performance of the ANFIS in signal prediction/system identification applications.

We will now consider the equalization of UWB systems using ANFIS in the following section.

4.4 Equalization of UWB Systems Using ANFIS

UWB is an emerging wireless technology that has recently gained much interest from the communication research industry. UWB systems possess unique characteristics and capabilities that make them suitable for short-range, high-speed wireless communications (Molisch 2005).

4.4.1 Introduction to UWB

UWB systems use signals that are based on repetitive transmissions of short pulses formed by using a single basic pulse shape. The transmitted signals have an extremely low power spectral density and occupy a very large bandwidth of several GHz. Thus UWB systems can operate with negligible interference to the existing radio systems. UWB can provide very high bit rate, low-cost, low-power wireless communication for a wide variety of systems: personal computer, TV, VCR, CD, DVD, and MP3 players (Molisch 2005, Algans et al. 2002).

UWB radars, which are mainly of interest for military applications, and UWB communications systems, which also have military applications, are nowadays mainly driven by commercial applications. UWB communications gained prominence with the groundbreaking work on impulse radio by Win and Scholtz in the 1990s (Qiu et al. 2005), and received a major boost by the 2002 decision of the U.S. frequency regulator (Federal Communications Commission, FCC) to allow unlicensed UWB operation.

As per FCC recommendations, UWB systems have the following characteristics:

- They have a *relative bandwidth that is larger than 25% of the carrier frequency and/or an absolute bandwidth more than* 500 MHz.

- They occupy a frequency band of 3.1 GHz to 10.6 GHz.

- The FCC has recently allocated 7.5 GHz of spectrum for unlicensed commercial UWB communication systems.

- Maximum radiated power is 75 *nW/MHz* (-41.32 *dBm/MHz*) (Molisch 2005).

The following are the significant merits of UWB:

1. Accurate position location and ranging, due to better time resolution.

2. No significant multipath fading due to better time resolution.

3. Multiple access due to wide transmission bandwidths.

4. Possibility of extremely high data rates.

5. Covert communications due to low transmission power operation.

6. Possible easier material penetration due to the presence of components at different frequencies.

4.4.2 Conventional Channel Models for UWB

Conventional wideband channel models discussed in Chapter 2 cannot be adapted as such to UWB due to the following reasons:

1. The signal conditioning problems associated with the wideband technology become more severe in the case of UWB. This includes CCI.

2. Rapid synchronization and acquisition is required for UWB.

3. Propagation models are more complex in multipath environments and do not allow for direct extension of narrow band.

4. The better time resolution of UWB results in different multipath components arriving at the receiver at different delays and angles, which creates a dynamic and extended CIR.

Typical environments and ranges for UWB are given in Table 4.3.

TABLE 4.3
Environments and Ranges for UWB Systems

Environment	Range
Indoor residential	1–30 m
Indoor office	1–100 m
Body Area Network (BAN)	0.1–2 m
Outdoor peer to peer	1–100 m
Outdoor base station scenario	1–300 m
Industrial environments	1–300 m
Emergency communications	1–50 m

TABLE 4.4
IEEE 802.15.3a Standard Summary Requirements

Parameter	Value
Bit Rate (PHY-SAP)	110 and 200 Mbps
Range	30 ft and 12 ft
Power Consumption	100 mW and 250 mW
BER	1×10^{-5}
Co-located Piconets	4
Interference Capability	Robust to IEEE Systems
Co-existence Capability	Reduced Interference to IEEE Systems

4.4.2.1 The Modified SV/IEEE 802.15.3a Model

The Modified Saleh–Valenzuela (SV) Model (SV/IEEE 802.15.3a) is commonly used to model UWB channels (Molisch 2005). The model developed by the IEEE 802.15.3a standardization group for UWB communications systems in order to compare standardization proposals for high data rate wireless PANs. Due to this purpose, the environments considered are office and residential indoor scenarios with a range of less than 10 m (Molisch 2005). Requirements of this model are summarized in Table 4.4.

4.4.2.2 The 802.15.4a Model for High Frequencies (4a HF)

The 802.15.4a standardization group has recently been developing a standard for UWB systems with low data rates and geolocation capabilities for sensor networks. The 802.15.3a models do not cover many of the ranges and environments envisioned for these applications, so that new models had to be developed. In addition, it decided to take into account several effects that were neglected in the 15.3a models. The resulting model for the 3–10 GHz range is a generalized SV model with parameters defined for residential indoor, office indoor, industrial, outdoor, and farm environments. For each of those environments, LOS and NLOS is distinguished, with the exception of farm environments, where only NLOS situations are modeled. The models are based on measurement campaigns, again with the exception of the farm environment, which is based on simulations only. Several of the underlying measurements do not cover the full 3–10 GHz range, restricting the validity range of the models (Molisch 2005).

4.4.2.3 The 802.15.4a Model for Low Frequencies (4a LF)

In addition to the 3–10 GHz range, the IEEE 802.15.4a group have also developed a model for the frequency range from 100 to 960 MHz (Molisch 2005). For this frequency range, only the office NLOS scenario is considered, since this is the only scenario where measurements are available. The chosen model is essentially the model of Cassioli et al. (2002), namely, a dense channel model with a single, exponentially decaying cluster. The decay constant is modeled as a deterministic variable that increases with distance as $(d/10$ m$)^{0.5} \times 40$ ns (note that this is a deviation from the original model of Cassioli et al. (2002). This equation gives the same delay spread as Cassioli, et al. (2002), at a 10 m distance. The distance exponent is chosen as a compromise between the results of Cassioli et al. (2002) (no distance dependence) and the results of Siwiak et al. (2003) that show a linear increase with distance. The first path has enhanced amplitude. The path gain follows a simple d^{-n} law, where n is the propagation (path gain) exponent.

4.4.2.4 Channel Covariance Matrix (CCM) Formulation

Shadow-fading fluctuations of the average received power are known to be log-normally distributed. Recently, for macrocell scenarios, the fluctuations in delay and angle spread are shown to behave similarly (Algans et al. 2002). The reason is that these quantities are sums of powers of individual sub-paths times the square of their corresponding delay times or angles. Since the powers are log-normally distributed and *sums of log-normal variables are (approximately) log-normal* (Beaulieu et al. 1995), this implies that angle spreads and delay spreads have log-normal distributions. This motivation of how angle spread and delay spread are log-normally distributed also suggests that they will be correlated with shadow fading and each other. Let us assume that $X_{1n}, X_{2n}, X_{3n}, \ldots$ are zero-mean, unit-variance Gaussian random variables, representing the signals received at base station n. Then we define:

$$\rho_{DA} = E[X_{1n}, X_{2n}] \qquad (4.19)$$

$$\rho_{DF} = E[X_{2n}, X_{3n}] \qquad (4.20)$$

$$\rho_{AF} = E[X_{3n}, X_{1n}] \qquad (4.21)$$

$$\zeta = E[X_{3n}, X_{3m}] \qquad (4.22)$$

In particular, $\sigma_{SF,n}$ (variance of shadow fading component w.r.t. to base station, n) is negatively correlated with $\sigma_{DS,n}$ (variance of *delay spread*) and $\sigma_{AS,n}$ (variance of *angle spread*), while the latter two have positive correlations with each other. It should be noted that this relationship does not hold for the angle spread at the mobile since the different paths with distinct angles do not necessarily lead to such pronounced differences in the delays. These correlations can be expressed in terms of a covariance matrix A, whose A_{ij} component represents the correlations between X_{in} and X_{jn}, with $i, j = 1,$ 2, 3. Note that the matrix A is symmetrical.

Measurements of cross-correlations of these parameters between different base stations are more difficult. In particular, only correlations between shadow-fading components have been adopted. These correlations are assumed to be the same between any two different base stations and are denoted by ζ. For simplicity and due to lack of further data, the cross-correlation matrix between the X_{in} triplet ($i = 1$, 2, 3) of different base stations is assumed to be given by the following matrix B.

$$A = \begin{bmatrix} 1 & \rho_{DA} & \rho_{DF} \\ \rho_{DA} & 1 & \rho_{AF} \\ \rho_{DF} & \rho_{AF} & 1 \end{bmatrix}, \quad B = \begin{bmatrix} 0 & 0 & 0 \\ 0 & 0 & 0 \\ 0 & 0 & \zeta \end{bmatrix} \quad (4.23)$$

4.4.2.5 Simulation of an ANFIS Equalizer for UWB Based on CCM

The following extended channel covariance matrix was used in the simulations.

$$A = \begin{bmatrix} 1 & 0.8 & -0.7 & 0.6 \\ 0.8 & 1 & -0.6 & 0.5 \\ -0.7 & -0.6 & 1 & 0.5 \\ 0.6 & 0.5 & 0.5 & 1 \end{bmatrix}$$

$$A' = \begin{bmatrix} 1+\alpha & 0.8+\beta & -0.7+\gamma & 0.6+\delta \\ 0.8+\beta & 1+\alpha & -0.6+\epsilon & 0.5+\varepsilon \\ -0.7+\gamma & -0.6+\epsilon & 1+\alpha & 0.5+\zeta \\ 0.6+\delta & 0.5+\varepsilon & 0.5+\zeta & 1+\alpha \end{bmatrix} \quad (4.24)$$

A' indicate the modified CCM corrupted by CCI and AWGN. We use an ANFIS with the following parameters in the equalizer:

- One-input one-output ANFIS.

- 20 rules.

- Gaussian membership functions.

- Maximum spread in CCM parameters is 0.5 (0.1:.01:0.5)

The structure of the ANFIS is given in Figure 4.13. The simulation results are given in Figure 4.14. They show that the ANFIS model is capable of estimating the CCM parameters with almost negligible error.

The MATLAB script for the simulation illustrated in Figure 4.14 is given below.

```
%%% MATLAB program to model the wideband channel
%%% Using  the covariance matrix.
%%% Last modified on  22-12-2012.
clc;clf;clear all;close all;
tic;
covm=[1 .8 -.7 .6;
          .8 1 -.6 .5;
          -.7 -.6 1 .5;
          .6 .5 .5 1];
%
```

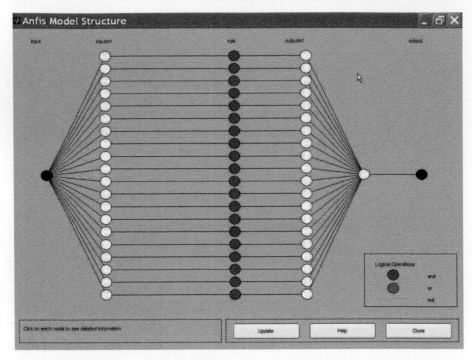

FIGURE 4.13 (See color insert.)
ANFIS Model Structure Used for UWB Channel Equalization: Number of
Inputs=1, Number of Outputs=1, Number of Rules=20, and Type of Membership Function- Gaussian.

```
for mc=1:4% for 4 MC simulations.
    mxerr=[];
    for spr=.1:.01:.5 %for  .. values of spread.
        x=covm(:);
        y(1)=x(1)+randn+spr;%first element with spread..
        y(2)=x(2)+randn+spr;%second element with spread..
        y(3)=x(3)+randn+spr;%third element  with spread..
        y(4)=x(4)+randn+spr;%fourth element with spread..
        y(5)=x(2); y(6)=x(1); y(11)=x(1); y(16)=x(1);
        y(7)=x(7)+randn+spr;
        y(8)=x(8)+randn+spr;
        y(9)=x(3);
        y(10)=x(10)+randn+spr;
        y(12)=x(12)+randn+spr;
        y(13)=x(4); y(14)=x(12); y(15)=x(12);
        y=y(1:16);
        trnData = [y'   x];
```

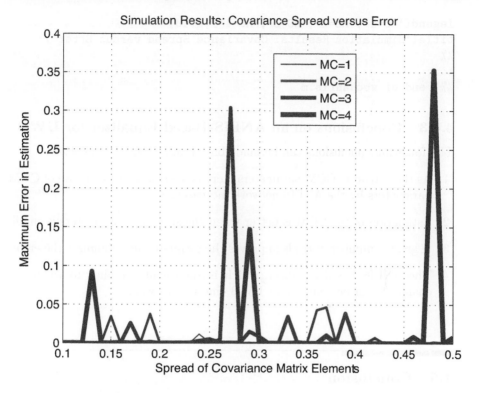

FIGURE 4.14 (See color insert.)
Results of Simulation–ANFIS Equalizer for UWB Channels: The results of
varying number of trials are indicated by varying line thicknesses. Note that
results improve as the number of trials increases.

```
    numMFs = 20;% No of Membership functions..
    mfType = 'gaussmf';
    epoch_n = 20;
    in_fismat = genfis1(trnData,numMFs,mfType);
    out_fismat = anfis(trnData,in_fismat,20);
    est_x=evalfis(y,out_fismat);
    err=(x-est_x);
    mxerr=[mxerr,max(err)];
     end;
    covm_eqln=reshape(est_x, 4,4)
    error_abs=reshape(err,4,4)
plot([.1:.01:.5],mxerr,'LineWidth',mc);hold on;
end;
xlabel('Spread of Covariance matrix elements');
ylabel('Maximum Error in Estimation ');
```

```
legend('MC=1','MC=2','MC=3','MC=4');grid;
title('Simulation Results: Covariance Spread versus Error');
%%
toc;
%%% end of wbchanfis.m
```

4.4.3   Conclusions on an ANFIS-Based Equalizer for UWB

The following conclusions can be made based on results of simulations:

1. As the spread in CCM parameters increases, error in the estimate of CCM parameters by the ANFIS network increases.

2. Estimation of the CCM is better when the spread in parameters is small.

3. A greater number of trials improves the estimate and minimizes the error.

4. The ANFIS based equalizer, which is successfully applied to wideband channels, can be adapted for UWB channels as well.

4.5   Conclusion

In this chapter, we considered an alternative solution to the nonlinear channel equalization problem. It was found that the ANFIS based equalizer performed nearly as well as the optimal *Bayesian* equalizer, as long as the SNR was greater than about 10 dB and the standard deviation of noise was low. Several ANFIS equalizer structures are considered, with varying numbers of inputs and membership functions. It was found that the BER versus SINR performance of all of them was almost the same. However, at low values of SNR, the ANFIS structure with more nodes performed slightly better. But as the number of nodes in the ANFIS structure was increased, convergence time was also increased, as evident from Table 4.2. The number of nodes in the ANFIS structure is a function of the number of inputs, membership functions, and outputs. The time for convergence increases as the number of inputs or membership functions increases.

It was also shown that equalizers based on ANFIS can be suitably adapted for UWB channels as well. A CCM formulation was used to model the UWB channel. It was shown that the estimate of the CCM was better when the *spread* in parameters was *small*.

Further Reading

J.S.R. Jang, ANFIS: Adaptive-Network-Based Fuzzy Inference System, *IEEE Transactions on Systems, Man, and Cybernetics*, Vol.23, No.3, pp.665–685, May/June 1993.

D. Williamson, R.A. Kennedy, and G.W. Pulford, Block Decision Feedback Equalization, *IEEE Transactions on Communications*, Vol.40, No.2, pp.255–264, February 1992.

B. Widrow and S.D. Stearns, *Adaptive Signal Processing*, Englewood Cliffs, NJ, Prentice-Hall, 1991.

G.D. Forney, Maximum-likelihood Sequence Estimation of Digital Sequences in the Presence of Inter-Symbol-Interference, *IEEE Transactions on Information Theory*, No.18, pp.363–378, 1972.

C.J .Lin and W.H. Ho, Blind Equalization Using Pseudo-Gaussian-Based Compensatory Neuro-Fuzzy Filters, *International Journal of Applied Science and Engineering*, Vol.2, No.1, pp.72–89, January 2004.

G.J. Gibson, S. Siu, and C.F.N. Cowan, The Application of Non-linear Structures to Reconstruction of Binary Signals, *IEEE Transactions on Signal Processing*, No.39, pp.1877–1884, 1991.

K.A. AlMashouq, and I.S. Reed, The Use of Neural Nets to Combine Equalization with Decoding for Severe Inter-Symbol-Interference Channels, *IEEE Transactions on Neural Networks*, No.5, pp.982–988, 1994.

S. Chen, G.J. Gibson, C.F.N. Cowan, and P.M. Grant, Reconstruction of Binary Signals Using Radial Basis Function Equalizer, *Signal Processing*, No.22, pp.77–93, 1991.

S. Chen, B. Mulgrew, and S. McLaughlin, Adaptive Bayesian Equalizer with Decision Feedback, *IEEE Transactions on Signal Processing*, Vol.41, pp.2918–2927, September 1993.

G. Kechriotis, and E.S. Manolakos, Using Recurrent Neural Networks for Adaptive Communication Channel Equalization, *IEEE Transactions on Neural Networks*, Vol.5, No.2, pp.267–278, 1994.

Qilian Liang and Jerry M. Mendel, Equalization of Nonlinear Time-Varying Channels Using Type-2 Fuzzy Adaptive Filters, *IEEE Transactions on Fuzzy Systems*, Vol.8, No.5, pp.551–563, October 2000.

Li-Xin Wang and Jerry M. Mendel, An RLS Fuzzy Adaptive Filter, with Application to Nonlinear Channel Equalization, *IEEE Transactions on Fuzzy Systems*, Vol.1, pp.895–900, August 1993.

C.T. Lin, and C.F. Juang, Adaptive Neural Fuzzy Filter and Its Applications, *IEEE Transactions on Systems, Man and Cybernetics (B)*, Vol.27, No.4, pp.640–656, April 1994.

S.K. Patra, and Bernard Mulgrew, Fuzzy Techniques for Adaptive Nonlinear Equalization, *Signal Processing*, No.80, pp.985–1000, 2000.

Lotfi A. Zadeh, Outline of a New Approach to the Analysis of Complex Systems and Decision Processes, *IEEE Transactions on Systems, Man and Cybernetics*, Vol.SMC-3, No.2, pp.28–44, January 1973.

Branimir B. Jovanovic, Irini S. Reljin, and Branimir D. Reljin, Modified AN-FIS Architecture–Improving Efficiency of ANFIS Technique, *Proceedings of the 7th Seminar on Neural Network Applications in Electrical Engineering*, NEURAL-2004, pp.215–220, September 23–25, 2004.

Andreas F. Molisch, Ultrawideband Propagation Channels—Theory, Measurement, and Modeling, *IEEE Transactions on Vehicluar Technology*, Vol. 54, No. 5, pp. 1528–1545, September 2005.

A. Algans, K. I. Pedersen, and P. E. Mogensen, Experimental Analysis of the Joint Statistical Properties of Azimuth Spread, Delay Spread, and Shadow Fading, *IEEE Journal on Selected Areas in Communication*, Vol. 20, No. 3, pp. 523–531, April 2002.

R.C. Qiu, H. Liu, and X. Shen, Ultra-Wideband for Multiple Access Communications, *IEEE Communication Magazine*, Vol. 43, pp. 80–87, 2005.

D. Cassioli, M.Z. Win, and A.F. Molisch, The Ultra-Wide Bandwidth IndoorChannel: From Statistical Model to Simulations, *IEEE Journal on Selected Areas in Communication*, pp. 1247–1257, 2002.

K. Siwiak, H. Bertoni, and S.M. Yano, Relation Between Multipath and Wave Propagation Attenuation, *Electronics Letters*, Vol.39, pp.142–143, January 2003.

N.C. Beaulieu, A.A. Abu-Dayya, and P.J. McLane, Estimating the Distribution of a Sum of Independent Lognormal Random Variables, *IEEE Transactions on Communications*, Vol.43, No.12, pp.2869–2873, December 1995.

T. Takagi and M. Sugeno, Fuzzy Identification on System and Its Applications to Modeling and Control, *IEEE Transactions on Systems, Man and Cybernetics*, Vol. 15, pp. 116–132, 1985.

5

Compensatory Neuro-Fuzzy Filter (CNFF)

In this chapter, we analyze the functioning of the channel equalizer based on the Compensatory Neuro-Fuzzy Filter (CNFF). It was stated in Chapter 4 that channel equalization is a nonlinear problem, so that a nonlinear solution is more appropriate. Moreover, for the mobile channel, *Blind Equalization* is the most preferred technique, due to the very special nature of the channel.

The rest of the chapter is organized as follows. In Section 5.1, we introduce the working principle of the CNFF. The channel equalizer based on the CNFF is introduced in Section 5.2. In Section 5.3, we consider the detailed structure of the CNFF (Lin and Ho 2003, 2004) for nonlinear channel equalization. Conclusions are made in Section 5.4.

5.1 Introduction

A fuzzy logic system (FLS) is unique in that it is able to simultaneously handle numerical data and linguistic knowledge. It is a nonlinear mapping of an input data (feature) vector into a scalar output, i.e., *it maps numbers into numbers*.

For many problems, two distinct forms of problem knowledge exist: the first one is objective knowledge, which is used all the time in engineering problem formulations (e.g., mathematical models), and the second one is subjective knowledge, which represents linguistic information that is usually impossible to quantify using traditional mathematics (e.g., rules, expert information, design requirements).

No standard methods exist for optimally transforming human knowledge or experience into the rule base and database of a fuzzy inference system. There is a need for effective methods for tuning the membership functions (MFs) so as to minimize the output error measure or maximize the performance index. The novel architecture of the Adaptive Network-Based Fuzzy Inference System (ANFIS) was discussed in greater detail in Chapter 4.

In this chapter, we discuss the CNFF, as an alternate technique for channel equalization. CNFF can be constructed by learning from training examples. It can be contrasted with traditional fuzzy logic control systems in their network structure and learning ability.

5.2　CNFF

The solution to the problem of channel equalization is targeted toward the removal of interference introduced by linear or nonlinear message corrupting mechanisms, so that the originally transmitted symbols can be recovered correctly at the receiver. In this section, we consider a CNFF based equalizer whose high performance makes it suitable for high-speed channel equalization. The compensatory fuzzy reasoning method is used in adaptive fuzzy operations that can make the fuzzy logic system more adaptive and effective. Besides, the pseudo-Gaussian membership function can provide the compensatory neuro-fuzzy filter which has a higher flexibility and approaches the optimized result more accurately. An online learning algorithm, which consists of structure learning and parameter learning, is proposed. Structure learning is based on the similarity measure of asymmetry Gaussian membership functions and parameter learning is based on the supervised gradient descent method. We apply the proposed CNFF to co-channel interference suppression (CCI) and Additive White Gaussian Noise (AWGN). Computer simulation results show that the bit error rate of the CNFF is close to the optimal equalizer (Lin and Ho 2004).

5.2.1　Outline of the CNFF

The CNFF is a four-layer structure (see Figure 5.1). Nodes at layer one are input nodes (linguistic nodes) which represent input linguistic variables. Layer four is the output layer. Nodes at layer two are *term nodes* which act as membership functions to represent the terms of the respective linguistic variable. Each node at layer three is a compensatory rule node which represents one fuzzy logic rule. Thus all layer-three nodes form a fuzzy rule base. The pseudo Gaussian (PG) membership functions ensure that the neuro-fuzzy filter has a higher flexibility and can approach the optimized result more accurately (Lin and Ho 2004). Besides, the compensatory fuzzy reasoning method is used in adaptive fuzzy operations that can make the fuzzy logic system more adaptive and effective. An online learning algorithm is used to construct the CNFF automatically. It consists of structure learning and parameter learning. The structure learning algorithm decides to add a new node which is satisfying the fuzzy partition of the input data. The similarity measure of asymmetry Gaussian membership functions is used to avoid the newly generated membership function being too similar to the existing one (Lin and Ho 2004). Back propagation learning is then used for tuning input/output membership functions.

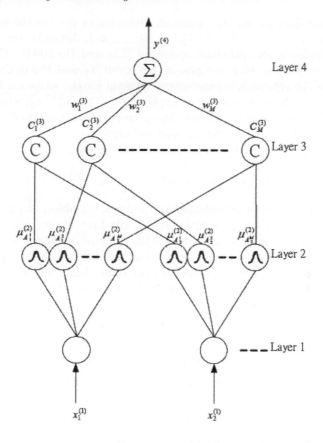

FIGURE 5.1
The Structure of the CNFF.

The learning method has the advantage that it does not require a human expert's assistance and it can converge quickly.

5.2.2 Details of Compensatory Operations

Zimmermann (Zimmermann and Zysno 1980) first defined the essence of compensatory operations. Zhang and Kandel (1998) have proposed more extensive compensatory operations based on the pessimistic operation and the optimistic operation. The pessimistic operation can map the inputs x_i to the pessimistic output by making a conservative decision for the pessimistic situation or even the worst case, for example, $p(x_1, x_2, \ldots, x_n) = MIN(x_1, x_2, \ldots, x_n)$ or $\prod_{i=1}^{n} x_i$. Actually, the $t-norm$ fuzzy operation is a pessimistic operation. The optimistic operation can map the inputs x_i to the optimistic output by making an optimistic decision for the optimistic situation or even in the best case, for example, $o(x_1, x_2, \ldots, x_n) = MAX(x_1, x_2, \ldots, x_n)$. Actually, the $t - conorm$ fuzzy operation is an optimistic operation (Lin and Ho 2004). The compensatory operation can map the pessimistic input x_1 and the optimistic input x_2 to make the relatively compromised decision for the situation between the worst case and the best case, for example, $c(x_1, x_2) = x_1^{1-\gamma} x_2^{\gamma}$, where $\gamma \in [0, 1]$ is called the compensatory degree. Many researchers have used the compensatory operation on fuzzy systems successfully.

The general fuzzy if–then rule is shown as follows.

$$R_j : IF \; x_1 \; is \; A_{1j} \; and \ldots and \; x_n \; is \; A_{nj} \; THEN \; y_b = b_j \quad (5.1)$$

where x_i, y_i are input dimensions and output variables; A_{ij} is a linguistic term of the precondition part with membership function $\mu_{A_{ij}}$; b_j is a constant consequent; n is an input dimension, i is the dimension index, $i = 1, \ldots, n$; n is the number of existing dimensions; j is the number of the rule, $j = 1, \ldots, R$; R is the number of existing rules. For an input fuzzy set A' in U, the jth fuzzy rule 5.1 can generate an output fuzzy set b_j' in v by using the $sup - dot$ composition

$$\mu_{b_j'} = \sup_{\underline{x} \in U} [\mu_{A_{1j} \times \ldots \times A_{nj} \to b_j}(\underline{x}, y) \circ \mu_{A'}(\underline{x})] \quad (5.2)$$

where $\underline{x} = (x_1, x_2, \ldots, x_n)$. The $\mu_{A_{1j} \times \ldots \times A_{nj}}(\underline{x})$ is defined in compensatory operation form 5.3 using the pessimistic operation 5.4 and the optimistic operation 5.5.

$$\mu_{A_{1j} \times \ldots \times A_{nj}}(\underline{x}) = (u_j)^{1-\gamma_j}(v_j)^{\gamma_j} \quad (5.3)$$

where $\gamma_j \in [0, 1]$ is a compensatory degree, and

$$u_j = \prod_{i=1}^{n} \mu_{A_{ij}}(x_i) \quad (5.4)$$

$$v_j = \left[\prod_{i=1}^{n} \mu_{A_{ij}}(x_i) \right]^{\frac{1}{n}} \quad (5.5)$$

After simplification, we can write

$$\mu_{A_{1j} \times \ldots \times A_{nj}}(\underline{x}) = \left[\prod_{i=1}^{n} \mu_{A_{ij}}(x_i) \right]^{1-\gamma_j+\frac{\gamma_j}{n}} \quad (5.6)$$

Since $\mu_A'(x) = 1$ for the singleton fuzzifier and $\mu_{b_j'}(x) = 1$, according to 5.2 we have

$$\mu_{b_j'}(y) = \left[\prod_{i=1}^{n} \mu_{A_{ij}}(x_i) \right]^{1-\gamma_j+\frac{\gamma_j}{n}} \quad (5.7)$$

5.3 Structure of CNFFs

A typical network consists of nodes with some finite number of fan-in connections from other nodes represented by weight values and fan-out connections to other nodes. Associated with the fan-in of a node is an integration function which combines information, activation, or evidence from other nodes, and provides the net input, i.e.,

$$net - input = f(z_1^{(k)}, z_2^{(k)}, \ldots, z_p^{(k)}; w_1^{(k)}, w_2^{(k)}, \ldots, w_p^{(k)}) \qquad (5.8)$$

where $z_i^{(k)}$ is the ith input to a node in layer k and $w_i^{(k)}$ is the weight of the associated link. The superscript in the above equation indicates the layer number (Lin and Ho 2004). This notation will also be used in the following equations. Each node also outputs an activation value as a function of its net input

$$output = a[f(.)] \qquad (5.9)$$

where $a(.)$ denotes the activation function. The CNFF is a network of four layers the (Figure 5.1), where the functions of the nodes in each layer are described as follows:

Layer 1: The nodes in this layer are input nodes (i.e., input-linguistic nodes), which represent input-linguistic variables and pass input signals to the next layer directly, i.e.,

$$f(x_i^{(1)}) = x_i^{(1)} \qquad (5.10)$$

and $a[f(.)] = f(.)$, where i is the input dimension index.

Layer 2: The nodes in this layer are term nodes that act as the PG membership function (Lin and Ho 2004). They can react on the terms of the respective input-linguistic variables. For the jth rule node

$$
\begin{aligned}
f(z_i^{(2)}) \;=\; & \exp\left(-\frac{(z_i^{(2)} - m_{ji})^2}{\sigma_{ji,-}^2}\right) U(z_i^{(2)}; -\infty, m_{ij}) \\
& + \exp\left(-\frac{(z_i^{(2)} - m_{ji})^2}{\sigma_{ji,+}^2}\right) U(z_i^{(2)}; m_{ij}, \infty) \qquad (5.11)
\end{aligned}
$$

and $a[f(.)] = f(.)$, where $U(z_i^{(2)}; a, b) = \begin{cases} 1 & if \ a \le z_i^{(2)} < b \\ 0 & otherwise. \end{cases}$

Layer 3: The nodes in this layer are compensatory fuzzy nodes. They represent the precondition part of the fuzzy logic rule, which can input the multiple incoming signals and output the product result. For the rule node

$$f(z_i^{(3)}) = \left[\prod_i^n z_i^{(3)}\right]^{1-\gamma+\frac{\gamma}{n}} \qquad (5.12)$$

and $a[f(.)] = f(.)$, where n is the dimension number.

Layer 4: The nodes in this layer are denoted by $\sum$. That is, the node receives the multiple incoming signals and outputs the result of summation. For the output

$$f(z_i^{(4)}) = \sum_{j=1}^{M} w_j^{(3)} z_i^{(4)} \tag{5.13}$$

and $a[f(.)] = f(.)$, where M is rule number; $w_j^{(3)}$ is the link weight.

5.3.1   Online Learning Algorithm

The online learning algorithm consists of the structure learning algorithm and the parameter learning algorithm. The structure learning algorithm is used to find proper fuzzy partitions in the input space and create fuzzy logic rules. An asymmetry similarity measure is proposed to avoid the newly generated membership function being too similar to the existing one. The parameter learning algorithm is the most general supervised learning scheme; it is used to adjust PG membership functions and compensatory operations in the precondition part, and modify the link weight in the consequent part. As a result, the parameter learning algorithm is based on the back propagation algorithm, which minimizes the cost function to approximate desired results. The procedure of the structure/parameter learning algorithm is through inputting the training pattern to learn successively (Lin and Ho 2004).

5.3.1.1   Structure Learning Algorithm

The proposition of the structure learning algorithm is to decide proper fuzzy partitions by the input patterns. The procedure is to find the proper fuzzy logic rules. However, the structure learning algorithm determines whether or not to add a new node in layer 2 via the input pattern data, and decides whether or not to add the associated fuzzy logic rule in layer 3. After the input pattern is entered in layer 2, the firing strength of the PG membership function will be obtained from Equation 5.11, which is used as the degree measure $\mu_{A_i^j}$. In layer 3, the firing strength of the fuzzy logic rule is obtained from Equation 5.12, which is used as the precondition part's degree measure

$$P = \prod_{j=1}^{M(i)} \mu_{A_i^j} \tag{5.14}$$

where i is the input dimension, $i = 1, \ldots, n$; j is the rule number, $j = 1, \ldots, M(t)$, $M(t)$ is the number of existing rules at time t.

To avoid the newly generated membership function being too similar to the existing one, the similarities between the new membership function and existing ones must be checked. If the new fuzzy rule is different from the existing fuzzy rule, the new fuzzy rule will be added in the CNFF. It can make the neural fuzzy inference system gain in performance. Therefore, we use a similarity measure of asymmetric Gaussian membership functions to estimate the rule's similarity degree (Lin and Ho 2004). Recall that for fuzzy sets A and B, their equivalence measure is calculated as

$$E(A,B) \underset{=}{\triangle} \frac{|A \cap B|}{|A \cup B|} \tag{5.15}$$

5.3.1.2 Parameter Learning Algorithm

After the structure network has been accordingly adjusted to the current training pattern, the network enters the parameter learning algorithm. The procedure of the parameter learning algorithm is to adjust the parameters of the CNFF optimally with the same training pattern. Back propagation is used for this supervised learning to find the output errors of the node in each layer and analyze the error to perform parameter adjustment (Lin and Ho 2004). The goal is to minimize the error function

$$E = \frac{1}{2} \left(y^d(t) - y(t) \right)^2 \tag{5.16}$$

where $y^d(t)$ is the desired output and $y(t)$ is the model output. Then the parameter learning algorithm based on backpropagation is as follows:

Assuming that w is the adjustable parameter in a node, the generally used learning rule is

$$w(t+1) = w(t) - \eta \left(\frac{\partial E}{\partial w} \right) \tag{5.17}$$

$$\frac{\partial E}{\partial w} = \frac{\partial E}{\partial f} \cdot \frac{\partial f}{\partial w}$$

$$= \frac{\partial E}{\partial a} \cdot \frac{\partial a}{\partial f} \cdot \frac{\partial f}{\partial w} \tag{5.18}$$

where η is the *learning rate*.

5.3.1.3 A Digital Communication System with AWGN and CCI

The discrete time model of a digital communication system with AWGN and CCI is shown in Figure 4.4. $H_0(z)$ is the desired channel and $H_i(z)$, $1 \leq i \leq n$, are the interfering co-channels. The impulse response of the channels and co-channels can be represented as

$$H_i(z) = \sum_{j=0}^{p_i} a_{ij} z^{-j} \tag{5.19}$$

Here p_i and a_{ij} are the length and tap weights of the ith channel impulse response. The transmitted sequences, $s_i(k)$; $1 \leq i \leq n$, are mutually independent and are taken from an independent, identically distributed data set with

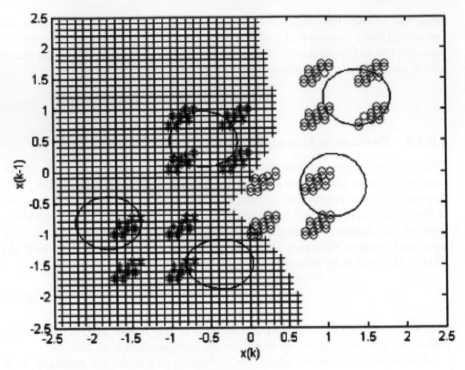

FIGURE 5.2 (See color insert.)
Decision Boundaries of the CNFF for $k = 50$. The boundary is marked by the mesh. The plot is a scattergram of the symbols received, $x(k - 1)$ and $x(k)$, at consecutive instances.

values $\{+1, -1\}$. The input to the equalizer forms the observation vector from the channel output (Lin and Ho 2004). Each of the components of this vector can be presented as

$$x(k) = \hat{x}(k) + \hat{x}_{co}(k) + e(k) \tag{5.20}$$

where $\hat{x}(k)$ is the desired received signal, and $\hat{x}_{co}(k)$ is the interfering signal. The noise $e(k)$ is assumed to be Gaussian with variance σ_e^2 and is uncorrelated with the data. The task of the equalizer is to estimate the delayed transmitted sequence $s_0(k - d)$ based on the channel observation vector $\mathbf{x}(k) = [x(k), x(k-1), \ldots, x(k - N + 1)]^T$. For the communication system with CCI and AWGN, the decision function of the *Bayesian Equalizer* is

$$f(x(k)) = \sum_{i=1}^{n_s} \sum_{m=1}^{n_{co}} \prod_{l=1}^{p-1} w_i \exp\left(-\frac{1}{2} \frac{[x(k - l) - \hat{x}_i(k - l) - \hat{x}_{co}^m(k - l)]^2}{\sigma_e^2 + \sigma_{co}^2} \right) \tag{5.21}$$

where $\hat{x}_{co}^m(k - l)$ is the lth element of the mth co-channel state (Lin and Ho 2004).

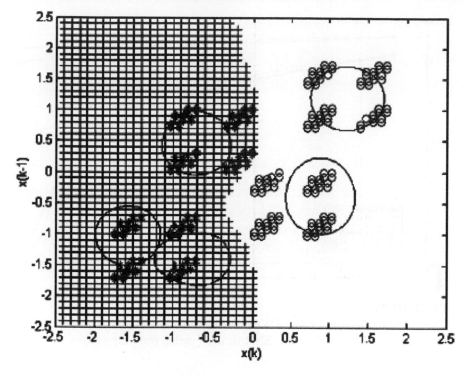

FIGURE 5.3 (See color insert.)
Decision Boundaries of the CNFF for $k = 100$. The boundary is marked by the mesh. The plot is a scattergram of the symbols received, $x(k - 1)$ and $x(k)$, at consecutive instances.

5.3.1.4 Channel Models and Simulation

The channel and the co-channels are characterized by their respective impulse responses.

$$H_{ch}(z) = 0.3482 + 0.8704z^{-1} + 0.3482z^{-2} \qquad (5.22)$$

$$H_{co1}(z) = \lambda(0.5 + 0.81z^{-1} + 0.31z^{-2}) \qquad (5.23)$$

The decision delay d is 1, and the input dimension n is 2. The initial parameters are chosen as $\eta = 0.01$ (learning rate), $E = 0.6$ (similarity threshold), and the standard deviation of interference noise due to the ith co-channel, σ_i is 0.5.

5.3.2 Simulation Results

The results of simulations (the decision boundaries) after the online training stopped at instances $k = 50$ and $k = 100$ are shown in Figures 5.2 and 5.3, respectively, with the number of rules generated being 5. To observe the actual

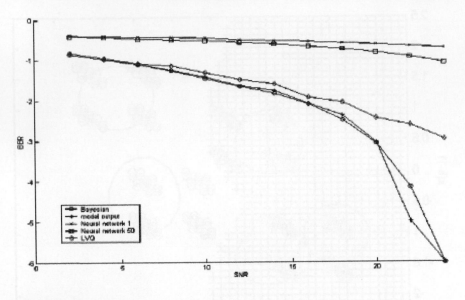

FIGURE 5.4
Comparison of BER of the CNFF with Bayesian/Neural Nets/LVQ: BER (varies from 10^{-6} to 1) versus SNR in dB (varies from 0 to 25).

BER, a realization of 10^6 points of the sequence $s(k)$ and $e(k)$ was used to test the BER of the trained network. The CNFF scheme is compared with Bayesian (optimal), Neural Network based and Linear Vector Quantization (LVQ) methods (Lin and Ho 2004). The BER curves for different SNRs are shown in Figure 5.4.

5.4　Conclusion

In this chapter, we considered an alternative solution to the nonlinear channel equalization problem. It was shown in Chapter 4 that the ANFIS based equalizer performs nearly as well as the optimal *Bayesian* equalizer, as long as the SNR is greater than about 10 dB and the standard deviation of noise is low.

We considered yet another equalizer structure for the equalization of nonlinear mobile cellular channels, which is CNFF. It was found that the bit error rate (BER) versus SNR performance of the CNFF is also close to other Neural Network based equalizers. Both the ANFIS and CNFF methods are nonlinear system approximators.

Further Reading

Cheng Jian Lin and Wen Hao Ho, A Pseudo-Gaussian-Based Compensatory Neuro-Fuzzy System, *Proceedings of the 12th IEEE International Conference on Fuzzy systems, FUZZ03*, Vol.1, pp.214–219, May 2003.

Cheng Jian Lin and Wen Hao Ho, Blind Equalization Using Pseudo-Gaussian-Based Compensatory Neuro-Fuzzy Filters, *International Journal of Applied Science and Engineering*, Vol.2, No.1, pp.72–89, January 2004.

H.J. Zimmermann, and P. Zysno, Latent Connective in Human Decision, *Fuzzy Sets and Systems*, No.4, pp.31–51, 1980.

Y.Q. Zhang, and A. Kandel, Compensatory Neuro-Fuzzy Systems with Fast Learning Algorithms, *IEEE Transactions on Neural Networks*, Vol.9, No.1, pp.83–105, January 1998.

6

Radial Basis Function Framework

In Chapters 3, 4, and 5, we have seen the realization of type-2 FAF, ANFIS, and CNFF-based equalizers, respectively. This chapter is dedicated to establishing a common link between them—that they can be brought under the generic framework of the Radial Basis Function (RBF) neural network. RBF based neural networks have been successfully used to solve many nonlinear problems, including that of adaptive equalization. In this context, we present three different adaptive fuzzy/neuro-fuzzy channel equalizers that closely fit into the framework of RBF neural network based systems. We consider the type-2 FAF based channel equalizer along with a CNFF and one based on an ANFIS as applied to mobile cellular channels. We establish that the three implementations of adaptive equalizers do fit into the generic framework of RBF based systems.

6.1 Introduction

The theory of RBF and their application to design channel equalizers is not new (Mulgrew 1996, Chen et al. 1993, Chandrakumar May 1998, September 1998, Erdogmus et al. 2001, Jang and Sun et al. 1993). Recently, there has been a new interest in the area, as evident from work by Xie and Leung et al. (2005). An important problem in data communications is that of channel equalization, that is, the removal of interference introduced by linear or nonlinear message corrupting mechanisms, so that the originally transmitted symbols can be recovered correctly at the receiver (Lin and Ho 2004). Channel equalization is an old problem, since the advent of telephone systems. Equalization is needed for Linear Time-Invariant (LTI) channels like the UTP cable or Coax, or the Linear Time Variant (LTV) channels like the radio channels used in mobile cellular telephony. The mobile cellular channel is time-variant due to *multipath fading*. The complexity of the equalizer increases as we move from the LTI channels to LTV channels. In the case of LTI channels like the UTP cable or Coax, the equalization is fairly less complex as the problem is basically that of system identification of an LTI system, and then obtaining the inverse system impulse response. In the case of LTV channels, on the other hand, the problem of equalization is very complex due to the time-variant nature of the system itself. Noise introduced in the channel is yet another issue, which makes the problem more severe. In the case of mobile cellular channels, it is shown by several authors that the channel is linear and time-varying and has either *Ricean fading* or *Rayleigh fading* characteristics (Tranter et al. 2004). The output SNR is affected by both Co-Channel Interference (CCI), which is present due to *frequency re-use*, and Adjacent Channel Interference (ACI), a contribution from the *spectral leakage* among frequencies used in adjacent channels in the cell (Rappaport 2003).

The rest of the chapter is organized as follows: In Section 6.2, we review the principles of RBFs. We discuss the implementation of the adaptive equalizers in Sections 6.3, 6.4, and 6.5. We make our observations and conclusions in Section 6.6.

6.2   RBF Neural Networks

Originally, RBF neural networks were developed for data interpolation in multi-dimensional space (Mulgrew 1996). Although the primary reason for using an equalizer on a communication channel has been to mitigate the effects of intersymbol interference, more recently it has been demonstrated that conventional equalizers can exploit the cyclostationary nature of the received signal and reduce the distortion due to both co-channel and adjacent channel interference. An RBF network can also be applied to this problem without the need to exploit the cyclostationary characteristics of the received signal. The structure of the RBF neural network is given in Figure 2.8. The use of the RBF has provided receivers with more controllable training characteristics than Multi-Layer Perceptron (MLP) receivers. However, the length of the training period is still too long for practical consideration. Blind equalization, in particular, is a demanding problem that currently receives a great deal of attention. While many techniques have been applied, RBF Bayesian methods have a unique contribution to play in this area, as they explicitly exploit the finite nature of the transmitted alphabet. This is unlike the techniques based on higher order statistics or the cyclostationary nature of the received signal, which are more complex. *RBFs neural networks can accommodate the channel non-linearity by effectively combining a large number of Gaussian basis functions.*

6.2.1   Review of Previous Work

The functional equivalence between RBF NNs and a simplified class of fuzzy inference systems was made by Jang and Sun (1993). This functional equivalence enables us to apply what has been discovered (learning rule, representational power, etc.) for one of the models to the other, and vice versa. It is of interest to observe that two models stemming from different origins turn out to be functionally equivalent. Though these two models are motivated from different origins (RBF networks from physiology and fuzzy inference systems from cognitive science), they share common characteristics not only in their operations on data, but also in their learning process to achieve the desired mappings. We show that under some minor restrictions, they are functionally equivalent; the learning algorithms and the theorem on representational power for one model can be applied to the other, and vice versa (Jang and Sun 1993).

The output of an RBF network can be computed in two ways. For the simpler one, as shown in Figure 2.8, the output is the weighted sum of the function value associated with each receptive field:

$$f(\vec{x}) = \sum_{i=1}^{N_r} f_i \, w_i = \sum_{i=1}^{N_r} f_i \, R_i(\vec{x}) \tag{6.1}$$

where f_i is the function value, or strength, of the ith receptive field.

The functional equivalence between an RBF network and a fuzzy inference system can be established if the following are true (Jang, and Sun 1993):

1. The number of receptive field units is equal to the number of fuzzy if–then rules.

2. The output of each fuzzy if–then rule is composed of a constant.

3. The membership functions within each rule are chosen as Gaussian functions with the same variance.

4. The *t-norm* operator used to compute each rule's firing strength is multiplication (or *product t-norm*).

5. Both the RBF NN and the fuzzy inference system under consideration use the same method (i.e., either *weighted average* or *weighted sum*) to derive their overall outputs.

6.2.1.1   Motivation for the Unified Framework

It is apparent by now that if we can establish a *unified framework* for adaptive equalizers based on neural networks and fuzzy logic, there will be several conveniences. There will be a synergical improvement in performance arising from combining the best features of both. We will tackle all three adaptive equalizers under consideration. Traditional adaptive algorithms for equalizers are based on the criterion of minimizing the mean square error between the desired filter output and the actual filter output, that is, these learning algorithms adjust the filter parameters to achieve a minimum of the criterion. Chen et al. (1993) have investigated the application of an RBF network to digital communications channel equalization an RBF. It is shown that the RBF network has an identical structure to the optimal Bayesian symbol-decision equalizer solution and, therefore, can be employed to implement the Bayesian equalizer. The training of an RBF network to realize the Bayesian equalization solution can be achieved efficiently using a simple and robust supervised clustering algorithm. *This represents a radically new approach to adaptive equalizer design.*

6.3   Type-2 FAF Equalizer

A channel with a more realistic equalizer order is used to study the performance of the RBF network under a variety of SNRs. The channel transfer function was given by Patra and Mulgrew (1998):

$$H(z) = 0.3482 + 0.8704z^{-1} + 0.3482z^{-2}. \tag{6.2}$$

Correct estimates of the channel order and the noise variance are assumed.

The type-2 FAF is realized using an unnormalized type-2 Takagi-Sugeno-Kang (TSK) fuzzy logic system (Liang and Mendel 2000, December 2000). A clustering method is used to adaptively design the parameters of the FAF. We used a transversal equalizer and decision feedback equalizer structures to eliminate the CCI. Simulation results show that the equalizers based on type-2 FAFs perform better than the nearest neighbor classifiers or the equalizers based on type-1 FAFs when the number of co-channels is much larger than 1, as described in Chapter 3. The statistical signal processing based approach (e.g., Bayesian decision rule) is based on a probability model (e.g., Gaussian distribution), whereas the FAF-based approach is model free. As noted in Mendel (2000), a shortcoming of model based statistical signal processing is the assumed probability model, for which model based statistical signal processing results will be good if the data agrees with the model, but may not be so good if the data does not.

A type-2 TSK FLS is described by fuzzy if–then rules which represent input–output relations of a system (Liang and Mendel 2000). The type-2 TSK FLS has a rule base of M rules, each having p antecedents, where the ith rule, R^i, is expressed as

$$R^i : IF \ x_1 \ is \ F_1^i \ and \ x_2 \ is \ F_2^i \ and \dots and \ x_p \ is \ F_p^i$$
$$THEN \ y^i = c_0^i + c_1^i \, x_1 + c_2^i \, x_2 + \dots + c_p^i \, x_p$$

in which $i = 1, 2, \dots, M$; $c_j^i (j = 0, 1, 2, \dots, p)$ are the consequent parameters; y^i is the output of the ith if–then rule; and, $F_k^i \ (k = 1, 2, \dots, p)$ are type-2 fuzzy sets.

6.3.0.1 A Simplified Mathematical Formulation for FAF-II

Equation 6.1 gives the output of an RBF NN in terms of the strengths of the receptive fields. One of the conditions for RBF NN and the FAF to be equivalent is that the number of receptive fields (N_r) in the RBF is equal to the number of fuzzy *if-then* rules (M) in the FAF. Now the membership function for the kth input is chosen as Gaussian with the variance $\sigma_k^{i^2}$ and mean m_k^i. It is also assumed that the *product t-norm* operator is used to compute the firing strength of each rule, and there are p inputs to the RBF NN and FAF. Then Equation 6.1 transforms to

$$
\begin{aligned}
f(\overrightarrow{x}) &= \sum_{i=1}^{N_r} f_i\, w_i = \sum_{i=1}^{N_r} f_i\, R_i(\overrightarrow{x}) \\
&= \sum_{i=1}^{M} y^i \times exp\left[-\frac{1}{2}\left(\frac{x_1 - m_1^i}{\sigma_1^i}\right)^2\right] \times \underbrace{exp\left[-\frac{1}{2}\left(\frac{x_2 - m_2^i}{\sigma_2^i}\right)^2\right] \dots \times}_{p\ terms} \\
&= \sum_{i=1}^{M} y^i \prod_{k=1}^{p} exp\left[-\frac{1}{2}\left(\frac{x_k - m_k^i}{\sigma_k^i}\right)^2\right] = y.
\end{aligned}
\tag{6.3}
$$

It can be observed that Equation 6.3 is identical to the output formula for an RBF network when Gaussian membership functions are used as the RBFs. This kind of RBF network has been applied to Bayesian equalization.

6.4 CNFF

The large computational complexity associated with the Viterbi algorithm and the poor performance of linear equalizers have led to the development of symbol-by-symbol equalizers using the Maximum a Posteriori probability (MAP) principle—Bayesian equalizers. The compensatory fuzzy reasoning method is used in adaptive fuzzy operations that can make the fuzzy logic system more adaptive and effective. Besides, the Pseudo-Gaussian (PG) membership function can provide the CNFF higher flexibility and it can approach the optimized result more accurately. An online learning algorithm, which consists of structure learning and parameter learning, is proposed. Structure learning is based on the similarity measure of asymmetry Gaussian membership functions and parameter learning is based on the supervised gradient descent method. We apply the proposed CNFF for CCI suppression and additive white Gaussian noise (AWGN) filtering. Computer simulation results show that the bit error rate of the CNFF is close to the optimal equalizer.

Bayesian equalizers have been approximated using nonlinear signal processing techniques like Artificial Neural Networks (ANN) (Gibson et al. 1991, AlMashouq and Reed 1994), RBFs (Chen et al. 1993), recurrent neural networks, and fuzzy filters (Sarwal and Srinath 1995, Wang and Mendel 1993). These new techniques provide the advantages of both good performance and low computational cost. Fuzzy filters are nonlinear filters that incorporate linguistic information in the form of if-then fuzzy rules. Fuzzy filters have been used for equalization due to their success in the related area of pattern classification (Liang and Mendel December 2000). Wang and Mendel (1993) have presented Fuzzy Basis Functions (FBFs) for channel equalization. Lin and Juang (1994) have developed the Artificial Neuro Fuzzy Filter (ANFF) and use it for equalization and noise reduction. This ANFF constructs its rule base in a dynamic way with the training samples. Patra and Mulgrew (2000) derived the close relationship between fuzzy equalizers and the equalizer based on the MAP. Liang and Mendel (2000) have developed type-2 fuzzy adaptive filters and demonstrated that they can implement the Bayesian equalizer.

The CNFF, which can be constructed by learning from training examples, can be contrasted with traditional fuzzy logic control systems in their network structure and learning ability. The CNFF is a four-layer structure (see Figure 5.1). Nodes at layer 1 are input nodes (linguistic nodes) which represent input linguistic variables. Layer 4 is the output layer. Nodes at layer 2 are term nodes which act as membership functions to represent the terms of the respective linguistic variable. Each node at layer 3 is a compensatory rule node, which, first explored systematically by Takagi and Sugeno, has found numerous practical applications in control, prediction, and inference (Jang 1993). However, there are some basic aspects of this approach which represent one fuzzy logic rule. Thus all the layer 3 nodes form a fuzzy rule base (Lin and Ho 2004). Besides, the compensatory fuzzy reasoning method is used in adaptive fuzzy operations that can make the fuzzy logic system more adaptive and effective.

The compensatory operation can map the pessimistic input x_1 and the optimistic input x_2 to make the relatively compromised decision for the situation between the worst case and the best case. For example, $c(x_1, x_2) = x_1^{1-\gamma} x_2^{\gamma}$, where $\gamma \in [0, 1]$ is called the compensatory degree (Lin and Ho 2004). Many researchers have used the compensatory operation for fuzzy systems successfully.

Nonlinear channel equalization is a technique used to combat some imperfect phenomenon in a high-speed channel (Lin and Ho 2004). The transmission input signal $s(k)$ is a sequence of statistically independent random binary symbols taking values $s(k) \in \{-1, 1\}$. The equalizer uses an input receiver signal vector $x(k) \in \mathcal{R}^m$, the m dimensional space, then the channel function can be described as

$$\hat{x}(k) = f\left[s(k), s(k-1), \ldots, s(k-N)\right]. \tag{6.4}$$

In general, f is a nonlinear function of the past transmitted signal, and the channels change slowly but significantly over time, so a nonlinear channel equalizer with adaptation ability is needed. At the receiving end, the observed signal $x(k)$ is the channel output $\hat{x}(k)$ corrupted by additive noise $e(k)$, that is, $x(k) = \hat{x}(k) + e(k)$. The noise source $e(k)$ is assumed to be zero mean white Gaussian with a variance of σ_e^2. The task of the equalizer is to reconstruct the transmitted signal $s(k-d)$ from the observed information sequence $x(k), x(k-1), \ldots, x(k-N+1)$ (where d and N denote the lag and order, respectively) such that greater speed and higher reliability can be achieved.

6.4.0.1 A Mathematical Formulation of CNFF

In a communication channel with AWGN (with zero mean and variance σ_e^2), but no CCI, the decision function of a Bayesian equalizer is

$$
\begin{aligned}
f(x(k)) &= \sum_{i=1}^{n_s} \prod_{l=1}^{p-1} f_i \exp\left(-\frac{1}{2} \frac{[x(k-l) - \hat{x}(k-l)]^2}{\sigma_e^2} \right) \\
&= \sum_{i=1}^{n_s} f_i \prod_{l=1}^{p-1} \exp\left(-\frac{1}{2} \frac{[x(k-l) - \hat{x}(k-l)]^2}{\sigma_e^2} \right) \quad (6.5)
\end{aligned}
$$

where f_i equals either $+1$ or -1 as determined by the channel state category. It is clear from Equation 6.5 that the CNFF output is similar to that obtained from an RBF NN, as given by Equation 6.1. With CCI, the numerator of Equation 6.5 gets modified as

$$
y = f(x(k)) = \sum_{i=1}^{n_s} f_i \sum_{m=1}^{n_{co}} \prod_{l=1}^{p-1} \exp\left(-\frac{1}{2} \frac{[x(k-l) - \hat{x}_i(k-l) - \hat{x}_{co}^m(k-l)]^2}{\sigma_e^2 + \sigma_{co}^2} \right)
$$

$$(6.6)$$

where $\hat{x}_{co}^m(k-l)$ is the lth element of the mth co-channel state.

6.5 ANFIS-Based Channel Equalizer

System modeling based on conventional mathematical tools (e.g., differential equations) is not well suited for dealing with ill-defined and uncertain systems. By contrast, a fuzzy inference system employing fuzzy *if–then* rules can model the qualitative aspects of human knowledge and reasoning processes without employing precise quantitative analyses. In the case of a Fuzzy Inference System (FIS),

1. No standard methods exist for transforming human knowledge or experience into the rule base and database of a fuzzy inference system.

2. There is a need for effective methods for tuning the membership functions (MFs), so as to minimize the output error measure or maximize the performance index.

Fuzzy if–then rules or fuzzy conditional statements are expressions of the form *IF A THEN B*, where A and B are labels of fuzzy sets, characterized by appropriate membership functions. Due to their concise form, fuzzy if–then rules are often employed to capture the imprecise modes of reasoning that play an essential role in the human ability to make decisions in an environment of uncertainty and imprecision. Another form of fuzzy if–then rule, proposed by Takagi and Sugeno, has fuzzy sets involved only in the premise (or antecedent) part. For example, by using Takagi and Sugeno's fuzzy if–then rule, we can describe the resistant force on a moving object as follows:

$$If\ velocity\ is\ high\ \ then\ force = k * (velocity)^2 \qquad (6.7)$$

where high in the premise part is a linguistic label characterized by an appropriate membership function. However, the consequent part is described by a nonfuzzy equation of the input variable. Both types of fuzzy if–then rules have been used extensively in both modeling and control. Through the use of linguistic labels and membership functions, a fuzzy if–then rule can easily capture the spirit of a *rule of thumb* used by humans (Jang 1993). The steps of *fuzzy reasoning* (inference operations upon fuzzy if–then rules) performed by fuzzy inference systems are described in Section 4.2.0.1 of Chapter 4.

6.5.0.1 A Mathematical Formulation of the ANFIS Equalizer

Several types of fuzzy reasoning have been proposed in the literature. Depending on the types of fuzzy reasoning and fuzzy if–then rules employed, most fuzzy inference systems can be classified into three types, as indicated in Section 4.2.0.1. We make use of a type-3 ANFIS with Gaussian membership functions for the channel depicted in Figure 4.4. The estimate of the channel output in this case can be expressed as

$$
\begin{aligned}
y &= \sum_{i=1}^{n_s} \overline{w}_i f_i = \frac{\sum_{i=1}^{n_s} w_i f_i}{\sum_{i=1}^{n_s} w_i} \\
&= \sum_{i=1}^{n_s} f_i \times \exp\left[-\frac{1}{2}\left(\frac{x_1 - m_{x_1}}{\sigma_{x_1}}\right)^2\right] \times \exp\left[-\frac{1}{2}\left(\frac{x_1 - x_{co_1}}{\sigma_{co_1}}\right)^2\right] \\
&\quad \times \exp\left[-\frac{1}{2}\left(\frac{x_2 - x_{co_2}}{\sigma_{co_2}}\right)^2\right] \cdots / \sum_{i=1}^{n_s} w_i \\
&= \sum_{i=1}^{n_s} f_i \times \frac{\exp\left[-\frac{1}{2}\left(\frac{x_i - m_{x_i}}{\sigma_{x_i}}\right)^2\right]}{\sum_{i=1}^{n_s} w_i} \times \prod_{k=1}^{p} \exp\left[-\frac{1}{2}\left(\frac{x_k - x_{co_k}}{\sigma_{co_k}}\right)^2\right] \quad (6.8)
\end{aligned}
$$

where $w_i = \exp\left[-\frac{1}{2}\left(\frac{x_i - m_{x_i}}{\sigma_{x_i}}\right)^2\right]$, $i = 1, 2, \ldots, n_s$. Note that $\overline{w}_i = \frac{w_i}{\sum_i w_i}$ is the normalized weight of f_i. Equation 6.8 shows that the ANFIS equalizer too can be brought under the generic framework of an RBF network. *It may be noted that the ANFIS accommodates nonlinearity by a convex combination of linear partitions.*

6.5.0.2 Simulations

Two comparative plots obtained after simulations are given in Figures 6.1 and 6.2.

The MATLAB script file used for the simulation illustrated in Figure 6.1 is appended below.

```
%%% MATLAB Script to simulate the performance of RBF NN
%%% and ANFIS-27..
%%% Here we design a exact radial basis network given inputs P
%%% and targets T.
%%  Last modified on 24-10-2012. The AWGN variance is changed
%%  between..
clc; clear all;close all;clf;
tic;
ns=64;% number of symbols.
nb=4;% number of bits per symbol
for mc=1:10% for 10 MC simulations.
    %for snr=15:2:23%for 5 values of SNR.
    n=1; snr=15;
    while snr<20.5
        %  for snr=15:1:20%for 6 values of SNR.
```

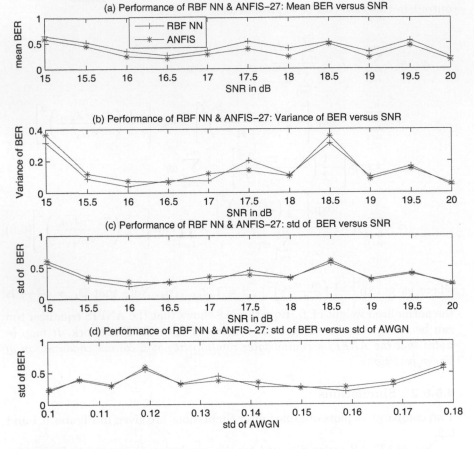

FIGURE 6.1 (See color insert.)
Performance of RBF NN and ANFIS-27: (a) Mean BER at output of the
equalizer versus SNR in dB, (b) Variance of BER versus SNR in dB, (c)
standard deviation of BER versus SNR in dB, and (d) standard deviation of
BER versus standard deviation of AWGN.

```
[x,b] = random_binary(ns,nb);
[x1,b1] = random_binary(ns,nb);
[x2,b2] = random_binary(ns,nb);
[x3,b3] = random_binary(ns,nb);
[x4,b4] = random_binary(ns,nb);
[x5,b5] = random_binary(ns,nb);
[x6,b6] = random_binary(ns,nb);
        z=x1+x2+x3+x4+x5+x6;
s=sqrt(var(x)/(10^(snr/10))); %std of AWGN
e=s*randn(1,ns*nb);% AWGN with std s.
```

```
y = x+0.18*z+e; % contribution from interfering channels
%is fixed..Std of Inter. is 0.18
y1=[y' y'];% 2 inputs..
%trnData = [y   x'];
P = y;
T = x;
%net = newrbe(P,T,2);%rbf spread is 2..
net = newrbe(P,T);%rbf spread is 0.2..
%Here the network is simulated for a new input.
Y = sim(net,P);
Y(Y<-0.6)=-1.0;
Y(Y>0.6)=1.0;
ec=(T~=Y);
ber1(mc,n)=sum(ec)/ns*nb;
%%
trnData = [y1 x'];
numMFs = 7;
mfType = 'gaussmf';
epoch_n = 20;
in_fismat = genfis1(trnData,numMFs,mfType);
out_fismat = anfis(trnData,in_fismat,20);
est_x=evalfis(y1,out_fismat);
est_x(est_x<-0.6)=-1.0;
est_x(est_x>0.6)=1.0;
ec=(x~=est_x');
ber2(mc,n)=sum(ec)/ns*nb;
n=n+1; snr=snr+0.5;
end; %end of while
end;%end of for
mBER1=mean(ber1);%mean BER of RBF NN
stBER1=std(ber1);% std of BER of RBF NN
vaBER1=var(ber1); % variance of BER--RBF NN
vaBER2=var(ber2) % variance of BER--ANFIS
mBER2=mean(ber2);%mean BER of ANFIS
stBER2=std(ber2);% std of BER of ANFIS
snr=[15:.5:20];
clf;
subplot(411), plot(snr,mBER1,'-+');hold on
plot(snr,mBER2,'-*'); hold on
xlabel('SNR in dB');
ylabel('mean BER');
title('(a)Performance of RBF NN & ANFIS--27: Mean BER versus SNR');
legend('RBF NN','ANFIS');
hold off;
subplot(412), plot(snr,vaBER1,'-+'); hold on
```

```
plot(snr,vaBER2,'-*');hold on
xlabel('SNR in dB');
ylabel('Variance of BER');
ch1=['(b)Performance of RBF NN & ANFIS--27: '];
ch2={' Variance of BER versus SNR'];
ch=strcat(ch1,ch2);
title(ch);
% legend('RBF NN','ANFIS');
hold off;
subplot(413), plot(snr,stBER1,'-+'); hold on
plot(snr,stBER2,'-*');hold on
xlabel('SNR in dB');
ylabel('std of  BER');
title('(c)Performance of RBF NN & ANFIS--27: std of BER versus SNR');
% legend('RBF NN','ANFIS');
hold off;
st=sqrt(var(x)./(10.^(snr/10)));
subplot(414),plot(st,stBER1,'-+'); hold on;
plot(st,stBER2,'-*'); hold on;
xlabel('std of AWGN');
ylabel('std of BER');
ch3=['(d)Performance of RBF NN & ANFIS--27:'];
ch4=[' std of BER versus std of AWGN'];
cx=strcat(ch1,ch2);
title(cx);
% legend('RBF NN','ANFIS');
hold off;
toc;
%%% end of rbfsnr27.m
```

The MATLAB script for the simulation illustrated in Figure 6.2 is appended below.

```
%%% MATLAB Script to simulate a RBF NN and ANFIS-25.
%%% Here we design a exact radial basis network given
%%  inputs P and targets T. Last modified on 24-10-2012.
%% The AWGN variance is changed between ..
clc; clear all;close all;clf;
tic;
ns=64;% number of symbols.
nb=4;% number of bits per symbol
for mc=1:10% for 10 MC simulations.
     %for snr=15:2:23%for 5 values of SNR.
     n=1; snr=15;
     while snr<20.5
          %  for snr=15:1:20%for 6 values of SNR.
```

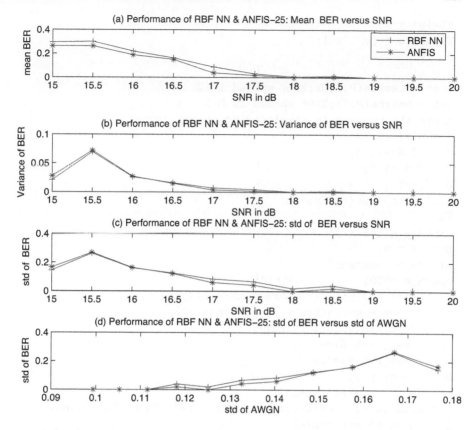

FIGURE 6.2 (See color insert.)
Performance of RBF NN and ANFIS-25: (a) Mean BER at output of the equalizer versus SNR in dB, (b) Variance of BER versus SNR in dB, (c) standard deviation of BER versus SNR in dB, and (d) standard deviation of BER versus standard deviation of AWGN.

```
[x,b] = random_binary(ns,nb);
[x1,b1] = random_binary(ns,nb);
[x2,b2] = random_binary(ns,nb);
[x3,b3] = random_binary(ns,nb);
[x4,b4] = random_binary(ns,nb);
[x5,b5] = random_binary(ns,nb);
[x6,b6] = random_binary(ns,nb);
        z=x1+x2+x3+x4+x5+x6;
s=sqrt(var(x)/(10^(snr/10))); %std of AWGN
e=s*randn(1,ns*nb);% AWGN with std s.
y = x+s*z+e;
```

```
y1=[y' y'];
%trnData = [y   x'];
P = y;
T = x;
%net = newrbe(P,T,2);%rbf spread is 2..
net = newrbe(P,T);%rbf spread is 0.2..
%Here the network is simulated for a new input.
Y = sim(net,P);
Y(Y<-0.6)=-1.0;
Y(Y>0.6)=1.0;
ec=(T~=Y);
ber1(mc,n)=sum(ec)/ns*nb;
%%
trnData = [y1 x'];
numMFs = 5;
mfType = 'gaussmf';
epoch_n = 20;
in_fismat = genfis1(trnData,numMFs,mfType);
out_fismat = anfis(trnData,in_fismat,20);
est_x=evalfis(y1,out_fismat);
est_x(est_x<-0.6)=-1.0;
est_x(est_x>0.6)=1.0;
ec=(x~=est_x');
ber2(mc,n)=sum(ec)/ns*nb;
% sinr(mc,n)=10*log10(var(x)./var(y));%%SINR of Ch1 output..
% intn(mc,n)=sum(s*z+e);
n=n+1; snr=snr+0.5;
end; %end of while
end;%end of for
mBER1=mean(ber1);%mean BER of RBF NN
stBER1=std(ber1);% std of BER of RBF NN
vaBER1=var(ber1); % variance of BER--RBF NN
vaBER2=var(ber2) % variance of BER--ANFIS
mBER2=mean(ber2);%mean BER of ANFIS
stBER2=std(ber2);% std of BER of ANFIS
snr=[15:.5:20];
% msinr=mean(sinr);
% stintn=std(intn);
clf;
subplot(411), plot(snr,mBER1,'-+');hold on
plot(snr,mBER2,'-*'); hold on
xlabel('SNR in dB');
ylabel('mean BER');
title('(a)Performance of RBF NN & ANFIS--25: SNR versus Average
          BER');
```

```
legend('RBF NN','ANFIS');
hold off;
subplot(412), plot(snr,vaBER1,'-+'); hold on
plot(snr,vaBER2,'-*');hold on
xlabel('SNR in dB');
ylabel('Variance of BER');
ch1=['(b)Performance of RBF NN & ANFIS--25: '];
ch2=[' SNR versus Variance of BER'];
ch=strcat(ch1,ch2);
title(ch);
% legend('RBF NN','ANFIS');
hold off;
subplot(413), plot(snr,stBER1,'-+'); hold on
plot(snr,stBER2,'-*');hold on
xlabel('SNR in dB');
ylabel('std of  BER');
title('(c)Performance of RBF NN & ANFIS--25: SNR versus std of
            BER');
% legend('RBF NN','ANFIS');
hold off;
st=sqrt(var(x)./(10.^(snr/10)));
subplot(414),plot(st,stBER1,'-+'); hold on;
plot(st,stBER2,'-*'); hold on;
xlabel('std of AWGN');
ylabel('std of BER');
ch3=['(d)Performance of RBF NN & ANFIS--25: '];
ch4=[' std of AWGN  versus std of BER'];
cx=strcat(ch3,ch4);
title(cx);
% legend('RBF NN','ANFIS');
hold off;
toc;
%%% end of rbfsinr.m
```

Figure 6.1 was obtained based on 100 Monte Carlo (MC) simulations. The inputs to the equalizers have both CCI and AWGN components, apart from the channel component. However, the standard deviation of the CCI signal is fixed at 0.18. In the first simulation, the variance of the AWGN is changed in accordance with the relation

$$\sigma_{\hat{n}} = \frac{\sigma_{\hat{x}}}{10^{SNR/10}} \tag{6.9}$$

where SNR = 15:0.5:20 dB, and $\sigma_{\hat{n}}$ and $\sigma_{\hat{x}}$ are variances of AWGN and the signal (the signal being a random binary waveform), respectively. Note that variances of all co-channels are set to the same value. Identical random signal inputs are given to a RBF NN based equalizer and an ANFIS–27 based equalizer, which has two inputs and 7 membership functions. The spread factor of the RBF NN was chosen as 1. In Figure 6.1(a), the output mean BER is plotted against the SNR of the channel output. In Figure 6.1(b) and (c), variance and standard deviation of BER are plotted against the SNR for the same simulation setup. In Figure 6.1(d), standard deviation of BER is plotted against standard deviation of AWGN. In Figure 6.2(a) to (d), the results of simulations on the same RBF NN (with spread 0.2) based equalizer and an ANFIS–25 based equalizer, under identical input conditions, are given. The slightly superior performance of the ANFIS–27 compared to the ANFIS–25 based equalizer as regards the mean BER performance is attributed to more rules (49) for the former, as against 25 for the latter. Therefore more precise system approximation is achieved by the ANFIS–27. But this is achieved at the cost of more time for convergence.

6.6   Conclusion

We have shown that all three Neuro-Fuzzy Equalizers discussed in previous sections fall into the generic framework of RBF NNs. Simulation results also indicate that the response of the RBF NN based equalizer is comparable to that of ANFIS based equalizer. However, as the shown by Figures 6.1 and 6.2, when the SNR is low, both ANFIS–25 and ANFIS–27 based equalizers slightly outperform the RBF NN, as far as the mean of BER is concerned. This is due to the more complex structure of the ANFIS. When the SNR is above 18.5 dB, both types of equalizers perform identically. Again, for low values of standard deviation of AWGN, the performance of the RBF NN equalizer is more or less the same as that of the ANFIS–25/ANFIS–27 equalizer, with respect to the standard deviation of BER. But, at high values of standard deviation of AWGN, the ANFIS equalizer performance is better. In the case of ANFIS–25 versus RBF NN, the performances are almost identical, as is evident from Figure 6.2(a) and (b). This shows that when the number of nodes is low (here it is 75), performance of the ANFIS is identical to that of the RBF NN.

The advantages in bringing all the three equalizers, viz., FAF-II, ANFIS, and CNFF-based equalizers into the generic framework of RBFs are the following:

1. The concept of the RBF NN is well known and hence optimization of the equalizer parameters is easy.

2. It is of great interest to investigate the performances of all three, and arrive at a particular solution which is most suited for a particular application scenario.

3. It is easier to arrive at the optimal *Bayesian equalizer* solution if we can bring in the generic framework.

Further Reading

B. Mulgrew, Applying Radial Basis Functions, *IEEE Signal Processing Magazine*, Vol.13, pp.50–65, March 1996.

Sheng Chen, Bernard Mulgrew, and Peter M.Grant, A Clustering Technique for Digital Communications Channel Equalization Using Radial Basis Function Networks, *IEEE Transactions on Neural Networks*, Vol.4, No.4, pp.570–579, July 1993.

P. Chandrakumar, P. Saratchandran, and N. Sundararajan, Non-Linear Channel Equalization Using Minimal Radial Basis Function Neural Networks, *Proceedings of the 1998 IEEE International Conference on Acoustics, Speech, and Signal Processing, ICASSP98*, Vol.6, pp.3373–3376, May 1998.

P. Chandrakumar, P. Saratchandran, and N. Sundararajan, Communication Channel Equalization Using Minimal Radial Basis Function Neural Networks, *Proceedings of the 1998 IEEE Signal Processing Society Workshop*, pp.477–485, August-September 1998.

D. Erdogmus, D. Rende, J.C. Principe, and T.F. Wong, Nonlinear Channel Equalization Using Multilayer Perceptrons with Information-Theoretic Criterion, *Proceedings of the International Workshop on Neural Networks for Signal Processing*, pp. 443–451, September 2001.

J.S.R. Jang, and C.T. Sun, Functional Equivalence Between Radial Basis Function Networks and Fuzzy Inference Systems, *IEEE Transactions on Neural Networks*, Vol.4, No.1, pp.156–159, January 1993.

Nan Xie and Henry Leung, Blind Equalization Using a Predictive Radial Basis Function Neural Network, *IEEE Transactions on Neural Networks*, Vol.16, No.3, pp.709–720, May 2005.

Cheng Jian Lin and Wen Hao Ho, Blind Equalization Using Pseudo-Gaussian-Based Compensatory Neuro-Fuzzy Filters, *International Journal of Applied Science and Engineering*, Vol.2, No.1, pp.72–89, January 2004.

William H. Tranter, K. Sam Shanmugan, Theodore S.Rappaport, and Kurt L. Kosbar, *Principles of Communication Systems Simulation with Wireless Applications*, Pearson Education, New Jersey 2004.

T.S. Rappaport, *Wireless Communications Principles and Practice*, Pearson Education, New Jersey 2003.

S.K.Patra and B. Mulgrew, Efficient Architecture for Bayesian Equalization Using Fuzzy Filters, *IEEE Transactions on Circuits and Systems II: Analog and Digital Signal Processing*, Vol.45, No.7, pp.812–820, July 1998.

Qilian Liang and Jerry M. Mendel, Equalization of Nonlinear Time-Varying Channels Using Type-2 Fuzzy Adaptive Filters, *IEEE Transactions on Fuzzy Systems*, Vol.8, No.5, pp.551–563, October 2000.

Qilian Liang and Jerry M. Mendel, Overcoming Time-Varying Co-Channel Interference Using Type-2 Fuzzy Adaptive Filters, *IEEE Transactions on Circuits and SystemsII: Analog and Digital Signal Processing*, Vol.47,No.12, pp.1419–1428, December 2000.

Jerry M. Mendel, Uncertainty, Fuzzy Logic, and Signal Processing, *Signal Processing*, Vol.80, No.6, pp.913–933, June 2000.

G.J. Gibson, S. Siu, and C.F.N. Cowan, The Application of Non-Linear Structures to Reconstruction of Binary Signals, *IEEE Transactions on Signal Processing*, No.39, pp.1877–1884, 1991.

K.A. AlMashouq, and I.S. Reed, The Use of Neural Nets to Combine Equalization with Decoding for Severe Inter-Symbol-Interference Channels, *IEEE Transactions on Neural Networks*, No.5, pp.982–988, 1994.

P. Sarwal and M.D. Srinath, A Fuzzy Logic System for Channel Equalization, *IEEE Transactions on Fuzzy Systems*, Vol.3, No.2, pp.246–249, May 1995.

Li-Xin Wang and Jerry M. Mendel, An RLS Fuzzy Adaptive Filter, with Application to Nonlinear Channel Equalization, *IEEE Transactions on Fuzzy Systems*, Vol.1,pp.895–900, August 1993.

C.T. Lin, and C.F. Juang, Adaptive Neural Fuzzy Filter and Its Applications, *IEEE Transactions on Systems, Man and Cybernetics (B)*,Vol.27, No.4, pp.640–656, April 1994.

S.K. Patra, and Bernard Mulgrew, Fuzzy Techniques for Adaptive Nonlinear Equalization, *Signal Processing*, No.80, pp.985–1000, 2000.

Jyh-Shing Roger Jang, ANFIS: Adaptive-Network-Based Fuzzy Inference System, *IEEE Transactions on Systems, Man, and Cybernetics*, Vol.23, No.3, pp.665–685, May/June 1993.

J.S.R. Jang, ANFIS: Adaptive-Network-Based Fuzzy Inference System, *IEEE Transactions on Systems, Man and Cybernetics*, Vol. 23, No. 3, pp. 665–685, 1993.

7

Modular Approach to Channel Equalization

The mobile cellular channel is generally considered to be Linear and Time-Variant (LTV). However, under limiting conditions, it can even be modeled as Linear and Time-Invariant (LTI). This approach is very convenient as it simplifies many problems associated with mobile cellular communication systems design such as equalizer design (which is in effect a system identification problem). In this conventional approach to modeling the mobile cellular channel as LTI or LTV, we assume that the transmitter subsystem, including channel encoder and modulator, is linear. Then only the notion of a linear time-invariant/variant channel holds good. But this is not the case often enough: Transmitter amplifiers are overdriven to their nonlinear region to have better power efficiency or to increase the transmitted power so as to increase the SNR (Lim et al. 1995, Lee and Gardner 2004). One way to model the transmitter nonlinearity is to shift it to the channel. The resulting nonlinear channel can easily be modeled and compensated by incorporating a *nonlinear* channel equalizer, which can be the *cascade of a nonlinear preprocessor filter and an LTV equalizer*. In such a case, the equalizer needs to tackle only the non-linearities inherent to the channel. We propose such a scheme in this chapter. Simulations indicate that this modular approach of shifting the non-linearity to the preprocessor from the inherently linear channel is indeed a good proposition.

7.1 Introduction

Mobile cellular channels and the equalizer design for them is a topic of intensive research activity. Most of the models used for mobile cellular channels assume a linear time-invariant/variant channel. Table 7.1 lists typical discrete LTI channel transfer functions. It may be noted that the discrete transfer functions listed in Table 7.1 are used in linear time-invariant channel modeling. The table considers only up to second order channels. The impulse response coefficients mentioned in Table 7.1 will be time varying about their nominal values in the case of LTV channels (Liang and Mendel 2000, December

TABLE 7.1
Transfer Functions of Linear Time-Invariant Channels

Channel	Transfer Function	Channel Type
$H_1(z)$	$0.5 + 1.0z^{-1}$	nonminimum phase
$H_2(z)$	$0.6 + 0.8z^{-1}$	nonminimum phase
$H_3(z)$	$1.0 + 0.2z^{-1}$	minimum phase
$H_4(z)$	$0.2682 + 0.9296z^{-1} + 0.2682z^{-2}$	mixed phase
$H_5(z)$	$0.5 + 0.81z^{-1} + 0.31z^{-2}$	mixed phase
$H_6(z)$	$0.3482 + 0.8704z^{-1} + 0.3482z^{-2}$	mixed phase
$H_7(z)$	$0.6963 + 0.6964z^{-1} + 0.1741z^{-2}$	mixed phase

2000). Modeling of an LTI system is fairly simple. Consequently, equalization of an LTI channel is also simple: it merely involves *finding the inverse system*, such that the overall impulse response function of the cascade of channel and equalizer results in a delayed unit impulse, $\delta(n - d)$. In the case of LTV channels, equalization is more complex, as the channel parameters constantly change with respect to time. As mentioned in Adali (1999), even in the case of a linear channel, a nonlinear equalizer is better due to the following reasons:

1. There exist several nonlinear equalizer techniques which are computationally efficient and provide good results in simulations.

2. The problem of equalization is inherently nonlinear and nonlinear equalizers converge faster.

3. In the presence of noise, and when the channel parameters are randomly varying, nonlinear modeling gives better results.

The use of large constellations provides bandwidth efficient modulation. Quadrature Amplitude Modulation (QAM) techniques have constellations, in which signal points are uniformly spread. Information is carried by both signal amplitude and phase; hence they are not constant envelopes. Thus, efficient nonlinear power amplifiers cannot be utilized in the transmitter without equalization in the receiver.

The rest of the chapter is organized as follows: In Section 7.2, we introduce the nonlinear channel model, which is most suitable for mobile cellular channels. Then in Section 7.3, we consider contemporary nonlinear equalizers based on RBF, Multi-Level Perceptrons (MLP), and Fuzzy Adaptive Filters (FAF) used for them. In Section 7.4, we introduce the modular approach. Simulation results for the proposed model are given in Section 7.5. We conclude the chapter in Section 7.6.

7.2   Nonlinear Channel Models

Practical power amplifiers introduce nonlinear distortion in the amplitude and the phase of the transmitted signal. The simple nonlinear model, described by Salehi, is widely used in developing methods to equalize nonlinear channels (Proakis and Salehi 2002). This model formulates the amplitude and phase distortion due to a nonlinear amplifier in the transmitter, using two simple two-parameter formulae. The input signal to the nonlinear channel can be written as

$$s(t) = a(t)cos[\omega_c t + \phi(t)] \tag{7.1}$$

Here, ω_c is the carrier frequency, $a(t)$ is the modulated amplitude, and $\phi(t)$ is the modulated phase. The amplitude and phase distortion are functions of the amplitude of the input signal, which are denoted by $A[a(t)]$ and $\Phi[a(t)]$, respectively. The output signal after the nonlinear channel is given by

$$r(t) = A[a(t)]cos\{\omega_c t + \phi(t) + \Phi[a(t)]\} \tag{7.2}$$

The model describes the distortions $A[a(t)]$ and $\Phi[a(t)]$ by the following functions (Proakis and Salehi 2002):

$$A[x] = \frac{\alpha_a x}{(1 + \beta_a x^2)} \tag{7.3}$$

$$\Phi[x] = \frac{\alpha_\phi x^2}{(1 + \beta_\phi x^2)} \tag{7.4}$$

Now we have

$$s_n(t) = \sum_{-\infty}^{\infty} a_n \exp(j\theta_n)p(t - nT) \tag{7.5}$$

Here the nth symbol interval is given by the amplitude and phase a_n and θ_n, T is the symbol interval, and $p(t)$ is the pulse waveform with duration T. The received signal, in complex baseband representation, is composed of the signal distorted by the nonlinear channel and a complex Gaussian noise with uncorrelated real and imaginary parts.

Linear equalizers that employ training sequences are often based on adaptive Finite Impulse Response (FIR) filters. They are easy to implement and track linear distortion in the channel fairly well provided that enough taps are used (using 50–100 taps is common). Some linear equalizers, such as a zero-forcing equalizer, may amplify channel noise (Qureshi 1985). As an alternative, nonlinear equalizers have the potential to compensate for all three sources of channel distortion. A common nonlinear equalizer is the Decision Feedback Equalizer (DFE).

7.3   Nonlinear Channel Equalizers

In this section we consider important nonlinear channel equalizers, which are contemporary, before we consider the modular approach.

7.3.1   Nonlinear Equalizers Based on RBF Neural Network

We now investigate various nonlinear equalizers based on Artificial Neural Networks (ANN) that are commonly used for equalization of cellular mobile channels.

Multi-layer feed-forward neural networks and RBF networks have been proposed recently to utilize the nonlinearity in the channel equalization problem. This is because ANN can easily perform nonlinear classifications and function associations.

A signal suffers from nonlinear, linear, and additive distortion when transmitted through a channel. Linear equalizers are commonly used in receivers to compensate for linear channel distortion. It is well known that nonlinear equalizers have the potential to compensate for all three sources of channel distortion, viz., channel nonlinearity, linear distortion, and additive white Gaussian noise (AWGN) (Lu and Evans 1999). Several authors have shown that nonlinear feedforward equalizers based on either MLP or RBF neural networks can outperform linear equalizers (Lu and Evans 1999). A reduced complexity neural network equalizer can be made by cascading an MLP and an RBF network. In simulation, the new MLP-RBF cascade equalizer outperformed MLP equalizers and RBF equalizers (Lu and Evans 1999).

Communication channel equalization using RBF neural networks is reported in the literature (Chandrakumar et al. 1998, September 1998). The economical network structure ensured by the Minimum Resource Allocation Network (MRAN) algorithm, which uses on-line learning, has the capability to grow and prune the RBF network's hidden neurons (Chandrakumar et al. September 1998). Compared to earlier methods, the MRAN algorithm does not have to estimate the channel order first, and fix the model parameters. The superiority of this method over existing methods is that separate channel order estimation is not necessary. The algorithm uses an Extended Kalman Filter (EKF) to determine the weight and width of each of the nodes. This is different from previous studies, where the width values have to be set to an estimate of the noise variance of the received data. To test the algorithm for nonlinear channels, the following nonlinear channel (Chen, et al. 1993) was chosen:

$$y(t) = x(t) + 0.2x^2(t) - 0.1x^3(t) + e(t) \qquad (7.6)$$

$$H(z) = \frac{X(z)}{S(z)} = 0.3482 + 0.8704z^{-1} + 0.3482z^{-2} \qquad (7.7)$$

The linear component $H(z)$ of the channel can be modeled as an FIR filter. The equalizer order is chosen as $m = 4$. In this example, the order of the linear part of the channel impulse response $n_h = 2$. Thus, there will be 64 desired states for the channel output $(2^{n_h+m} = 64)$. The decision delay was set to one $(\tau = l)$. The MRAN algorithm is used to train the equalizer with 500 data samples at an SNR of 20 dB. The structure of the conventional channel equalizer scheme is depicted in Figure 7.1. The proposed novel scheme of adding a prepocessor filter is illustrated in Figure 7.2.

The MRAN algorithm using RBF neural networks is seen to be well suited for channel equalization problems. Its ability to build up a network based on certain parameters is seen to have an advantage over other methods, as it can be used for on-line training of the data for equalization. The algorithm's performance is evaluated by using it to build up an equalization network for two channels (linear and nonlinear). The resulting networks are then tested by comparing their bit error rate performance to that of the Bayesian equalizer. The results show that the networks obtained are comparable in performance to Bayesian equalizers when suitable training parameters are selected.

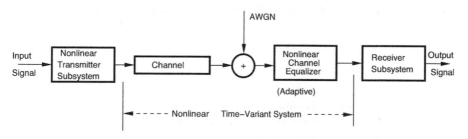

FIGURE 7.1
Conventional Equalizer Scheme.

The MATLAB script used for the simulation illustrated in Figure 7.3 is given below.

```
%%% MATLAB Script to simulate the Preprocessor scheme..
%%% Non-Linear Channel Eqlr ANFIS
%%% Last modified on 24-10-2012.
clc;clear all;close all;clf;
tic;
%%Training data..
tc = gauspuls('cutoff',50e3,0.6,[],-40);
%sets the cutoff level of Gauss Pulse as -40dB..
t = -tc : 1e-6 : tc;
yi = gauspuls(t,50e3,0.6);% Generates One Guassian pulse..
x=(ones(10,1)*yi)';%Replicates 10 Gaussian pulses..
x=x(:)';
xch=(x+0.2*x.^2-0.1*x.^3+0.16*randn(1,length(x)));
```

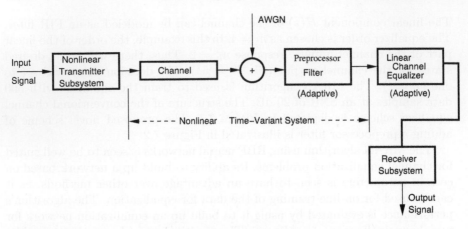

FIGURE 7.2
Linear Equalizer with Preprocessor Filter—Proposed Scheme.

```
% xch=(x+0.4*x.^2-0.2*x.^3+0.16*randn(1,length(x)));
y=[xch',xch'];
trnData = [y x'];
numMFs =7;
mfType = 'gaussmf';
epoch_n = 20;
in_fismat = genfis1(trnData,numMFs,mfType);
out_fismat = anfis(trnData,in_fismat,20);
%%%Cheching Data..same as input..
est_x=evalfis(y,out_fismat);
%est_y(est_y>1.0)=1.0;
%est_y(est_y<-1.0)=-1.0;
t=[1:length(x)];
subplot(411),plot(t,x,'LineWidth',2);grid;
axis([0 750 -1.5 1.5]);
xlabel('Time, t'); ylabel('Undistorted Input');
title('Modulator Output');
subplot(412),plot(t,xch,'LineWidth',2);grid;
axis([0 750 -1.5 1.5]); % axis([1 128 -1.5 1.5]);
xlabel('Time, t'); ylabel('Channel  Output');
title('Input to Prefilter');
subplot(413),plot(t,est_x,'LineWidth',2);grid;
axis([0 750 -1.5 1.5]); % axis([1 128 -2 2]);
xlabel('Time, t'); ylabel('Preprocessor Output');
title('ANFIS-27 Output');
b=[0.3482 0.8704 0.3482];
a=1;
```

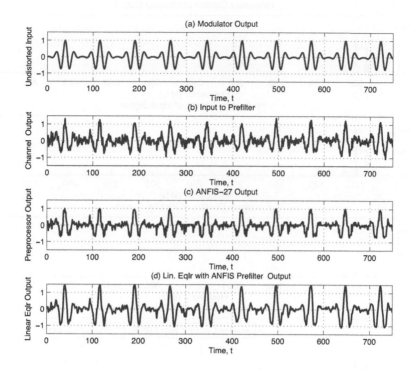

FIGURE 7.3 (See color insert.)
Simulation Results of Preprocessor Scheme with ANFIS Prefilter, Showing the *Time Domain Responses*. The input signal $x(t)$ is a Gaussian pulse train; the output of the channel is given by $y(t) = x(t) + 0.2x^2(t) - 0.1x^3(t) + \eta(t)$. Note that waveforms (a) and (d) are closer to each other than (a) and (c).

```
yf=filter(b,a,est_x);
subplot(414),plot(t,yf,'LineWidth',2); grid;
axis([0 750 -1.5 1.5]); %axis([1 128 -2 2]);
xlabel('Time, t'); ylabel('Linear Eqlr Output');
title('Lin.Eqlr with ANFIS Prefilter  Output');
toc;
%%% end of gauanfis.m
```

The MATLAB script used for the spectra illustrated in Figure 7.4 is appended below.

```
%%% MATLAB Script to plot the Spectra of
%%% Non-Linear Channel EQlr using ANFIS-27
%%% Last modified on 24-10-2012.
```

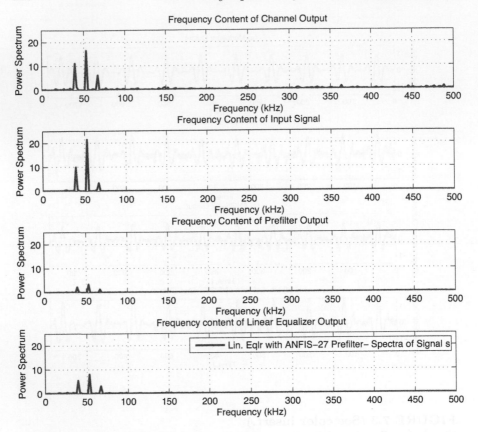

FIGURE 7.4 (See color insert.)
Spectra of Signals in Figure 7.3

```
%%% Input is a Gaussian Pulse Train..
clc;clear all; close all;clf;
tic;
%%Training data..
tc = gauspuls('cutoff',50e3,0.6,[],-40);
%sets the cutoff level of Gauss Pulse as -40dB..
t = -tc : 1e-6 : tc;
yi = gauspuls(t,50e3,0.6);% Generates One Guassian pulse..
x=(ones(10,1)*yi)';%Replicates 10 Gaussian pulses..
x=x(:);
xch=(x+0.2*x.^2-0.1*x.^3+0.44*randn(length(x),1));
%Noise Variance is 0.24.
y=[xch,xch];
trnData = [y x];
numMFs =7;
```

```
mfType = 'gaussmf';
epoch_n = 20;
in_fismat = genfis1(trnData,numMFs,mfType);
out_fismat = anfis(trnData,in_fismat,20);
%%%Cheching Data..same as input..
Y=evalfis(y,out_fismat);
% Ouput of ANFIS--27 Nonlinear Prefilter..
Xch= fft(xch,512);
PXch = Xch.* conj(Xch) / 512; % power spectrum
f = 1000*(0:256)/512;
subplot(411),plot(f,PXch(1:257),'LineWidth',2);
axis([0 500 0 25]);grid;
title('Frequency Content of Channel Output')
xlabel('Frequency (kHz)'); ylabel('Power Spectrum');
X= fft(x,512);
PX = X.* conj(X) / 512; % power spectrum
subplot(412),plot(f,PX(1:257),'LineWidth',2);
axis([0 500 0 25]);grid;
title('Frequency Content of Input Signal')
xlabel('Frequency (kHz)');ylabel('Power Spectrum');
Xh= fft(Y,512);
PXh = Xh.* conj(Xh) / 512; % power spectrum
subplot(413),plot(f,PXh(1:257),'LineWidth',2);
axis([0 500 0 25]);grid;
title('Frequency Content of Prefilter Output')
xlabel('Frequency (kHz)');ylabel('Power  Spectrum');
b=[0.3482 0.8704 0.3482]; %Linear phase Chl.
% b=[0.6963 0.6964 0.1741]; % Minimum phase
% b=[0.5 0.81 0.31];
a=1;
yf=filter(b,a,Y);
Yf=fft(yf,512);
PYf=Yf.*conj(Yf)/512;%Power spectrum.
subplot(414),plot(f,PYf(1:257),'LineWidth',2);
axis([0 500 0 25]);  grid;
title('Frequency content of Linear Equalizer Output')
xlabel('Frequency (kHz)');ylabel('Power Spectrum');
ch1=['Lin.Eqlr with ANFIS--27 Prefilter'];
ch2=['- Spectra of Signals'];
ch=strcat(ch1,ch2);
legend(ch);
toc;
%%% end of gauanfsp.m
```

The MATLAB program used for the above simulation, illustrated in Figure 7.5, is appended below

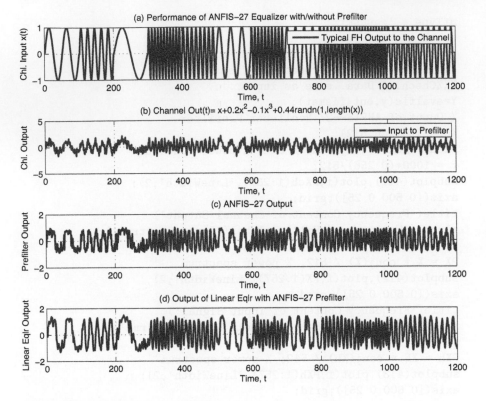

FIGURE 7.5 (See color insert.)

Simulation Results with ANFIS Prefilter for a Frequency-Hopped (FH) Carrier. The signal $x(t)$ is an FH Carrier; the output $y(t) = x(t) + 0.2x^2(t) - 0.1x^3(t) + \eta(t)$. Note that waveforms (a) and (d) are closer to each other than (a) and (c).

```
%%% MATLAB script to simulate a
%%% Non-Linear Channel Eqlr using ANFIS..
%%% The input to the ANFIS is a FH signal
%%% generated using the fh.m script.
%%% Last modified on 24-10-2012.
clc;clear all;close all;clf;
tic;
%%Training data..
rand('state',0) % sets the seed to 0.
[ignore,h] = sort(randn(1,12));% random permutations ...
t=linspace(0,1,100);
kf=1e4;%Freq Multiplication factor..
x=[];
for i=1:12
```

```
    xn=sin(2*pi*kf*h(i)*t);
    x=[x,xn];
end;
%%%Non-Linear Channel Output
xch=(x+0.2*x.^2-0.1*x.^3+0.44*randn(1,length(x)));
% xch=(x+0.4*x.^2-0.2*x.^3+0.16*randn(1,length(x)));
y=[xch',xch'];
trnData = [y x'];
numMFs =7;
mfType = 'gaussmf';
epoch_n = 20;
in_fismat = genfis1(trnData,numMFs,mfType);
out_fismat = anfis(trnData,in_fismat,20);
%%%Cheching Data..same as input..
est_x=evalfis(y,out_fismat);
t=[1:length(x)];
subplot(411),plot(t,x,'LineWidth',2);grid;
%axis([0 750 -1.5 1.5]);
xlabel('Time, t'); ylabel('Chl. Input x(t)');
ch1=['Performance of ANFIS-27 Equalizer'];
ch2=[' with/without Prefilter'];
ch=strcat(ch1,ch2);
title(ch);
legend('Typical FH Output to the Channel');
subplot(412),plot(t,xch,'LineWidth',2);grid;
%axis([0 750 -1.5 1.5]);
xlabel('Time, t'); ylabel('Chl. Output');
ch3=['Channel Out(t)= x+0.2x^2-0.1x^3+'];
ch4=['0.44randn(1,length(x))'];
cx=strcat(ch3,ch4);
title(cx);
legend('Input to Prefilter');
subplot(413),plot(t,est_x,'LineWidth',2);grid;
%axis([0 750 -1.5 1.5]);
xlabel('Time, t'); ylabel('Prefilter Output');
title('ANFIS-27 Output');
b=[0.3482 0.8704 0.3482];
a=1;
yf=filter(b,a,est_x);
subplot(414),plot(t,yf,'LineWidth',2); grid;
%axis([0 750 -1.5 1.5]);
xlabel('Time, t'); ylabel('Linear Eqlr Output');
title('Output of Linear Eqlr with ANFIS-27 Prefilter');
toc;
%%% end of fhanfis.m
```

The MATLAB script to generate an FH signal is appended below.

```
%%% MATLAB script to generate a Frequency Hopped signal
%%% to test the Equalizer performance..
%%% Uses the randn function, with 'seed' set to 0..
%%  Last modified on 24-10-2012.
clear all;close all; clc;
rand('state',0) % sets the seed to 0.
[ignore,h] = sort(rand(1,12));
t=linspace(0,1,1000);
kf=1e3;
x=[];
for i=1:12
    xn=sin(2*pi*kf*h(i)*t);
    x=[x,xn];
end;
plot(x);
%%% end of fh.m
```

The MATLAB script used for the above simulation, resulting in the spectra illustrated in Figure 7.6, is given below.

```
%%% MATLAB Script to simulate the
%%% Spectral Analysis of Non-Linear Channel EQlr
%%% The input to the ANFIS is a FH signal
%%% generated using the fh.m script.
%%  Last modified on 24-10-2012.
clc;clear all;close all;clf;
tic;
%%Training data..
rand('state',0) % sets the seed to 0.
[ignore,h] = sort(randn(1,12)); % random permutations ...
t=linspace(0,1,100);
kf=1e4;%Freq Multiplication factor..
x=[];
for i=1:12
    xn=sin(2*pi*kf*h(i)*t);
    x=[x,xn];
end;
%%%Non-Linear Channel Output
xch=(x+0.2*x.^2-0.1*x.^3+0.44*randn(1,length(x)));
% xch=(x+0.4*x.^2-0.2*x.^3+0.16*randn(1,length(x)));
y=[xch',xch'];
trnData = [y x'];
numMFs =5;
mfType = 'gaussmf';
```

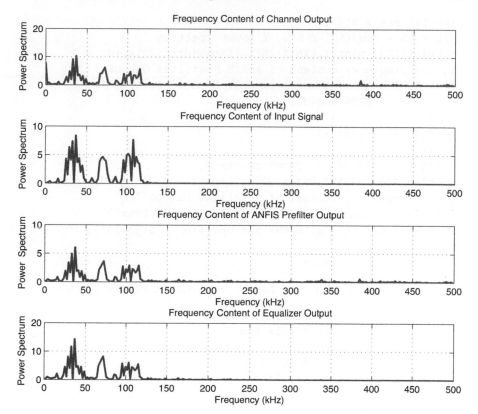

FIGURE 7.6 (See color insert.)
Spectra of Signals in Figure 7.5.

```
epoch_n = 20;
in_fismat = genfis1(trnData,numMFs,mfType);
out_fismat = anfis(trnData,in_fismat,20);
%%%Cheching Data..same as input..
est_x=evalfis(y,out_fismat);
Xch= fft(xch,512);
PXch = Xch.* conj(Xch) / 512; % power spectrum
f = 1000*(0:256)/512;
subplot(411),plot(f,PXch(1:257),'LineWidth',2);grid;
title('Frequency content of Channel Output')
xlabel('Frequency (kHz)'); ylabel('Power Spectrum');
X= fft(x,512);
PX = X.* conj(X) / 512; % power spectrum
subplot(412),plot(f,PX(1:257),'LineWidth',2);grid;
title('Frequency content of Input Signal')
xlabel('Frequency (kHz)');ylabel('Power Spectrum');
```

```
Xh= fft(est_x,512);
PXh = Xh.* conj(Xh) / 512; % power spectrum
subplot(413),plot(f,PXh(1:257),'LineWidth',2);grid;
title('Frequency content of ANFIS Prefilter Output')
xlabel('Frequency (kHz)');ylabel('Power  Spectrum');
b=[0.3482 0.8704 0.3482];
a=1;
yf=filter(b,a,est_x);
Yf=fft(yf,512);
PYf=Yf.*conj(Yf)/512;%Power spectrum.
subplot(414),plot(f,PYf(1:257),'LineWidth',2); grid;
title('Frequency content of Equalizer Output')
xlabel('Frequency (kHz)');ylabel('Power Spectrum');
% legend('Lin.Eqlr with RBF NN- Spectra of Signals');
toc;
%%% end of fhanfissp.m
```

The MATLAB script used for the simulation illustrated in Figure 7.7 is given below.

```
%%% MATLAB Script to simulate a
%%% Non-Linear Channel Eqlr based on RBF NN
%% Last modified on 24-10-2012.
clc;clear all;close all;clf;
tic;
%%Training data..
tc = gauspuls('cutoff',50e3,0.6,[],-40);
%sets the cutoff level of Gauss Pulse as -40dB..
t = -tc : 1e-6 : tc;
yi = gauspuls(t,50e3,0.6);% Generates One Guassian pulse..
x=(ones(10,1)*yi)';%Replicates 10 Gaussian pulses..
x=x(:)';
xch=(x+0.2*x.^2-0.1*x.^3+0.24*randn(1,length(x)));
y=xch;
P = y;
T = x;
net = newrbe(P,T,2);
%Here the network is simulated for a new input.
Y = sim(net,P);
subplot(411),plot(x,'LineWidth',2);
axis([0 750 -1.5 1.5]);grid;
xlabel('Time, t'); ylabel('Undistorted Input');
title('Training Data');
subplot(412),plot(xch,'LineWidth',2);
axis([0 750 -1.5 1.5]);grid;
xlabel('Time, t'); ylabel('Channel  Output');
```

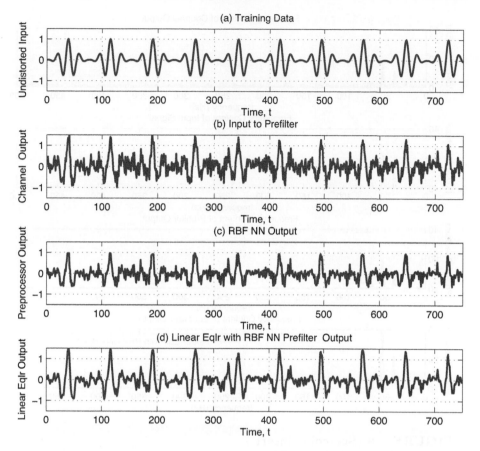

FIGURE 7.7 (See color insert.)
The signal $x(t)$ is a Gaussian pulse train; the output is given by $y(t) = x(t) + 0.2x^2(t) - 0.1x^3(t) + \eta(t)$. Note that waveforms (a) and (d) are closer to each other than (a) and (c).

```
title('Input to Prefilter');
subplot(413),plot(Y,'LineWidth',2);
axis([0 750 -1.5 1.5]);grid;
xlabel('Time, t'); ylabel('Preprocessor Output');
title('RBF NN Output');
b=[0.3482 0.8704 0.3482];
a=1;
yf=filter(b,a,Y);
subplot(414),plot(yf,'LineWidth',2);
grid;axis([0 750 -1.5 1.5]);
xlabel('Time, t'); ylabel('Linear Eqlr Output');
```

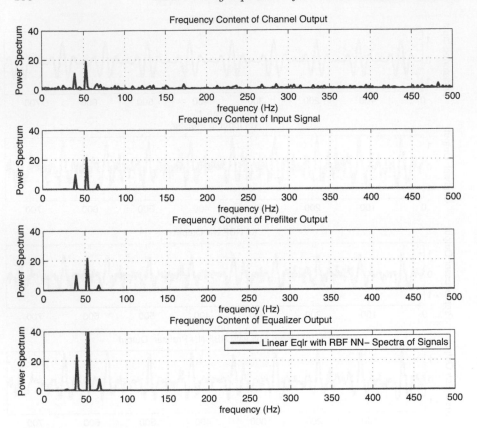

FIGURE 7.8 (See color insert.)
Spectra of Signals in Figure 7.7.

```
title('Linear Eqlr with RBF NN Prefilter  Output');
toc;
%%% end of gaurbf.m
```

The MATLAB script to obtain the spectra shown in Figure 7.8 is appended below.

```
%%% MATLAB Script to obtain the spectra of
%%% Non-Linear Channel EQlr with RBF NN Prefilter..
%% Last modified on 24-10-2012.
clc;clear all;close all;clf;
tic;
%%Training data..
tc = gauspuls('cutoff',50e3,0.6,[],-40);
%sets the cutoff level of Gauss Pulse as -40dB..
t = -tc : 1e-6 : tc;
```

```
yi = gauspuls(t,50e3,0.6);% Generates One Guassian pulse..
x=(ones(10,1)*yi)';%Replicates 10 Gaussian pulses..
x=x(:);
xch=(x+0.2*x.^2-0.1*x.^3+0.8*randn(length(x),1));
y=[xch];
P = y;
T = x;
net = newrbe(P,T,2);
%Here the network is simulated for a new input.
Y = sim(net,P);
Xch= fft(xch,512);
PXch = Xch.* conj(Xch) / 512; % power spectrum
f = 1000*(0:256)/512;
subplot(411),plot(f,PXch(1:257),'LineWidth',2);
axis([0 500 0 40]);grid;
title('Frequency content of Channel Output');
xlabel('frequency (Hz)'); ylabel('Power Spectrum');
X= fft(x,512);
PX = X.* conj(X) / 512; % power spectrum
subplot(412),plot(f,PX(1:257),'LineWidth',2);
axis([0 500 0 40]);grid;
title('Frequency content of Input Signal');
xlabel('frequency (Hz)');ylabel('Power Spectrum');
Xh= fft(Y,512);
PXh = Xh.* conj(Xh) / 512; % power spectrum
subplot(413),plot(f,PXh(1:257),'LineWidth',2);
axis([0 500 0 40]);grid;
title('Frequency content of Prefilter Output');
xlabel('frequency (Hz)');ylabel('Power  Spectrum');
b=[0.3482 0.8704 0.3482];
a=1;
yf=filter(b,a,Y);
Yf=fft(yf,512);
PYf=Yf.*conj(Yf)/512;%Power spectrum.
subplot(414),plot(f,PYf(1:257),'LineWidth',2);
axis([0 500 0 40]);  grid;
title('Frequency content of Equalizer Output');
xlabel('frequency (Hz)');ylabel('Power Spectrum');
legend('Linear Eqlr with RBF NN- Spectra of Signals');
toc;
%%% end of gaurbfsp.m
```

The MATLAB script to generate the above simulation, illustrated in Figure 7.9, is appended below.

```
%%% MATLAB Script to simulate a
```

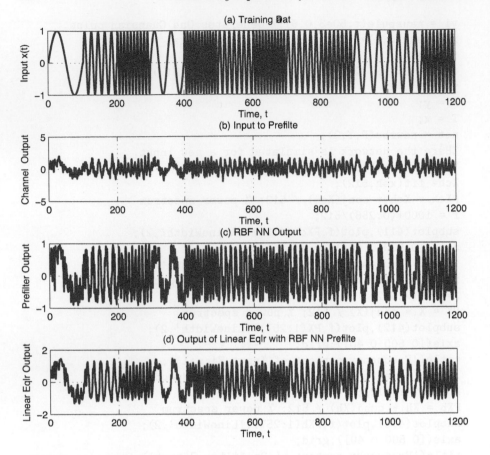

FIGURE 7.9 (See color insert.)
Simulation Results: RBF NN Prefilter with FH Carrier. Note that waveforms
(a) and (d) are closer to each other than (a) and (c).

```
%%% Non-Linear Channel Eqlr based on RBF NN Prefilter..
%%% The input to the RBF NN (with spread 2) is a FH signal
%%% generated using the fh.m script.
%% Last modified on 24-10-2012.
clc;clear all;close all;clf;
tic;
%%Training data..
rand('state',0) % sets the seed to 0.
[ignore,h] = sort(randn(1,12)); % random permutations ...
t=linspace(0,1,100);
kf=1e4;%Freq Multiplication factor..
x=[];
```

```
for i=1:12
    xn=sin(2*pi*kf*h(i)*t);
    x=[x,xn];
end;
%%%Non-Linear Channel Output
xch=(x+0.2*x.^2-0.1*x.^3+0.44*randn(1,length(x)));
%%Training data..
y=xch;
P = y;
T = x;
net = newrbe(P,T);
%Here the network is simulated for a new input.
Y = sim(net,P);
subplot(411),plot(x,'LineWidth',2); grid;
xlabel('Time, t'); ylabel('Input x(t)');
title('Training Data');
subplot(412),plot(xch,'LineWidth',2); grid;
xlabel('Time, t'); ylabel('Channel  Output');
title('Input to Prefilter');
subplot(413),plot(Y,'LineWidth',2); grid;
xlabel('Time, t'); ylabel('Prefilter Output');
title('RBF NN Output');
b=[0.3482 0.8704 0.3482];
a=1;
yf=filter(b,a,Y);
subplot(414),plot(yf,'LineWidth',2);
grid;%axis([0 750 -1.5 1.5]);
xlabel('Time, t'); ylabel('Linear Eqlr Output');
title('Output of Linear Eqlr with RBF NN Prefilter');
toc;
%%% end of fhrbf.m
```

The MATLAB script to obtain the spectra shown in Figure 7.10 is appended below.

```
%%% MATLAB script to generate the
%%% Spectral Analysis of Non-Linear Channel EQlr
%% with RBF NN Prefilter..
% The input to the RBF NN is a FH signal
% generated using the fh.m script.
%Last modified on 24-10-2012.
clc;clear all; close all; clf;
tic;
%%Training data..
rand('state',0) % sets the seed to 0.
[ignore,h] = sort(randn(1,12)); % random permutations ...
```

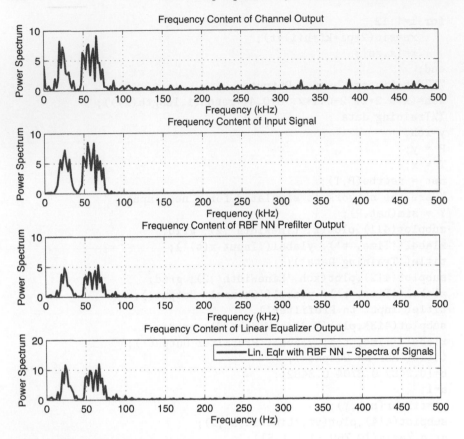

FIGURE 7.10 (See color insert.)
Spectra of Signals in Figure 7.9.

```
t=linspace(0,1,100);
kf=1e4;%Freq Multiplication factor..
x=[];
for i=1:12
    xn=sin(2*pi*kf*h(i)*t);
    x=[x,xn];
end;
%%%Non-Linear Channel Output
xch=(x+0.2*x.^2-0.1*x.^3+0.44*randn(1,length(x)));
% xch=(x+0.4*x.^2-0.2*x.^3+0.16*randn(1,length(x)));
y=[xch];
P = y;
T = x;
net = newrbe(P,T,1);
```

```
%Here the network is simulated for a new input.
Y = sim(net,P);
Xch= fft(xch,512);
PXch = Xch.* conj(Xch) / 512; % power spectrum
f = 1000*(0:256)/512;
subplot(411),plot(f,PXch(1:257),'LineWidth',2);
%axis([0 500 0 25]);
grid;
title('Frequency Content of Channel Output')
xlabel('Frequency (kHz)'); ylabel('Power Spectrum');
X= fft(x,512);
PX = X.* conj(X) / 512; % power spectrum
subplot(412),plot(f,PX(1:257),'LineWidth',2);
%axis([0 500 0 25]);
grid;
title('Frequency Content of Input Signal')
xlabel('Frequency (kHz)');ylabel('Power Spectrum');
Xh= fft(Y,512);
PXh = Xh.* conj(Xh) / 512; % power spectrum
subplot(413),plot(f,PXh(1:257),'LineWidth',2);
%axis([0 500 0 25]);
grid;
title('Frequency Content of RBF NN Prefilter Output')
xlabel('Frequency (kHz)');ylabel('Power  Spectrum');
b=[0.3482 0.8704 0.3482];
a=1;
yf=filter(b,a,Y);
Yf=fft(yf,512);
PYf=Yf.*conj(Yf)/512;%Power spectrum.
subplot(414),plot(f,PYf(1:257),'LineWidth',2);  grid;
title('Frequency Content of Linear Equalizer Output')
xlabel('Frequency (Hz)');ylabel('Power Spectrum');
legend('Lin.Eqlr with RBF NN- Spectra of Signals');
toc;
%%% end of fhrbfsp.m
```

7.3.2 Nonlinear Equalizers Based on MLPs

The idea of using MLPs has existed in the literature with successful examples of improved performance over linear equalizers. The MLP equalizer consists of two MLPs operating in parallel. One of them, MLP1, is trained to learn the mapping from the amplitude of the transmitted symbol, $|S|$, to the amplitude of the received signal, $|R|$, where S and R are phasors, obtained from the signals by integrating over one symbol duration and scaling down by the symbol duration. Assigning the input-output variables in this manner also helps the

MLP to avoid modeling the noise in the received signal. The other, MLP2, is trained to learn the mapping from $|R|$ to the phase shift introduced by the nonlinear channel, where the desired output is given by the phase difference $\angle R - \angle S$ between the received and transmitted symbols (Erdogmus et al. 2001).

The two MLPs are trained both with a single hidden layer with 6 neurons and a linear output neuron using the entropy minimization algorithm. The training set consisted of 360 symbols. The variance of the discrete-time noise is adjusted to achieve a predetermined SNR at the equalizer input. SNR here represents the ratio of average bit energy to noise Power Spectral Density (PSD). For each SNR value MLPs are trained and tested independently. In training the MLPs, steepest ascent for information potential is used. A dynamic step size, whose value increases when the update yields a better performance and decreases when the performance degrades, is utilized. It is observed that the weights of MLPs converged to the optimal solution in about 20–30 iterations, for all SNR values, with an initial step size of 1. It is observed that these MLPs converged in 100 iterations starting with the same step size. Upon completion of the training process, the equalizers are tested for Bit Error Rate (BER) using appropriate noise levels and sufficiently long test bit sequences (Erdogmus et al. 2001).

Some remarkable properties of the proposed equalizer are its computational simplicity, due to the small size of MLPs that can achieve good performance, efficient extraction of information from a small number of training samples, due to the information-theoretic optimality criterion, and the robustness to the radial component of the additive channel noise.

7.3.3   Nonlinear Equalizers Based on FAFs

The most commonly used recent fuzzy models are type-1 FAF (FAF–I) as proposed by Patra and Mulgrew (2000), and an improved version by Liang and Mendel (2000). A still different approach is to use the ANFIS (Jang 1993). There are some very recent innovations in blind channel equalization using predictive RBF neural networks (Xie and Leung 2005). Jang has established the functional equivalence between fuzzy inference systems and RBF neural networks (Jang, and Sun 1993).

7.4   A Modular Approach for Nonlinear Channel Equalizers

As shown in Figure 7.1, conventional linear as well as nonlinear channel equalizers are cascaded to a linear time-variant (or nonlinear time-variant) channel to combat Inter-Symbol-Interference (ISI). In the case of a linear transmitter stage followed by an LTV channel, an LTV or NLTV channel equalizer would suffice. As mentioned in Section 7.2, the proposed system model incorporating a preprocessor filter that takes care of the channel nonlinearities arising due to transmitter subsystem is shown in Figure 7.2. The proposed paradigm is based on a divide and conquer rule. To realize the preprocessor filter, we used an ANFIS–27 with the following parameters:

Number of Rules=49 (131 nodes); Membership Function Type: Gaussian; and Number of Epochs=20.

As a second method, we also use an RBF NN with spread 1 as the preprocessor filter. The results of the simulations are discussed in the following section.

7.5   Simulation Results

For the simulations, we used the input–output relation given in Equation 7.6. In the first simulation, a Gaussian pulse train was used as the signal, $x(t)$. The standard deviation of the AWGN at the channel is taken as 0.2. The input signal, the input to the preprocessor, the output of the preprocessor filter, and the output of the linear equalizer are given in Figure 7.3(a) to (d). The corresponding spectra of the respective signals are shown in Figure 7.4(a) to (d). In the second simulation, we used an FH carrier as the signal, $x(t)$. The results are given in Figures 7.5 and 7.6. It is evident from the plots that the ANFIS based prefilter is effective in eliminating the higher order nonlinearities. Rest assured, the *LTV Channel Equalizer* may be able to estimate the input sequence more closely.

In an entirely different set of simulations, we used an RBF NN to act as the preprocessor filter. Simulation results for identical input vectors are as shown in Figure 7.7(a) to (d) and 7.8. It can be seen from Figures 7.7 and 7.8, that performance is almost comparable. Similarly, for an FH carrier, the response of the RBF NN based prefilter and the output of the linear equalizer are given in Figure 7.9. The corresponding spectra of signals are given in Figure 7.10.

7.6   Conclusion

We have shown that preprocessor filters based on the ANFIS and RBF NN are effective tools in combating nonlinearities introduced by the *transmitter subsystem* and the *channel*. Even though we have used ANFIS and RBF NN to realize the preprocessor filter, any suitable adaptive filter structure described in previous sections and generic FIR structures can also be used. Also, it is necessary to process the output of the prefilter in the equalizer, using any of the methods discussed in previous sections. We can conclude that the merits for going for a preprocessor filter are

1. The preprocessor filter can take care of the nonlinearities introduced by the channel and the transmitter. Since the transmitter power amplifier is preceding the channel, the channel nonlinearity is to be taken care of first.

2. With the preprocessor filter, the role of the equalizer is getting simplified, as the former removes the nonlinearities introduced by the channel and the transmitter Power Amplifier (PA). Thus the combination of the preprocessor filter and equalizer has the merit of simplicity in design as well as an improvement in performance.

Further Reading

Yong Hoon Lim, Yong Soo Cho, Il Whan Cha, and Dae Hoe Youn, An Adaptive Nonlinear Prefilter for Compensation of Distortion in Nonlinear Systems, *IEEE Transactions on Signal Processing*, Vol.46, No.6, pp.1726–1730, June 1995.

K.C. Lee, and P. Gardner, A Combined Neural Network and Fuzzy Systems Based Adaptive Digital Predistortion for RF Power Amplifier Linearization, *Proceedings of the 47th IEEE International Midwest Symposium on Circuits and Systems*, Vol.3, pp.61–64, July 2004.

Qilian Liang and Jerry M. Mendel, Equalization of Nonlinear Time-Varying Channels Using Type-2 Fuzzy Adaptive Filters, *IEEE Transactions on Fuzzy Systems*, Vol.8, No.5, pp.551–563, October 2000.

Qilian Liang and Jerry M. Mendel, Overcoming Time-Varying Co-Channel Interference Using Type-2 Fuzzy Adaptive Filters, *IEEE Transactions on Circuits and Systems II: Analog and Digital Signal Processing*, Vol.47, No.12, pp.1419–1428, December 2000.

Tulay Adali, Why a Nonlinear Solution for a Linear Problem? *Proceedings of IEEE International Workshop on Neural Networks for Signal Processing*, pp. 157–165, 1999.

John G. Proakis and Masoud Salehi, *Communication Systems Engineering*, 2nd edition, Pearson Education, New Jersey 2002.

S.U.H. Qureshi, Adaptive Equalization, *Proceedings of the IEEE*, Vol.73, pp.1349–1387, September 1985.

Biao Lu and Brian L. Evans, Channel Equalization by Feedforward Neural Networks, *Proceedings of IEEE International Symposium on Circuits and Systems*, Orlando, FL, Vol.5, pp.587–590, May 30-June 2, 1999.

P. Chandrakumar, P. Saratchandran, and N. Sundararajan, Non-Linear Channel Equalization Using Minimal Radial Basis Function Neural Networks, *Proceedings of the 1998 IEEE International Conference on Acoustics, Speech, and Signal Processing, ICASSP98*, Vol.6, pp.3373–3376, May 1998.

P. Chandrakumar, P. Saratchandran, and N. Sundararajan, Communication Channel Equalization Using Minimal Radial Basis Function Neural Networks, *Proceedings of the 1998 IEEE Signal Processing Society Workshop*, pp.477–485, August-September 1998.

Sheng Chen, Bernard Mulgrew, and Peter M.Grant, A Clustering Technique for Digital Communications Channel Equalization Using Radial Basis Function Networks, *IEEE Transactions on Neural Networks*, Vol.4, No.4, pp.570–579, July 1993.

D. Erdogmus, D. Rende, J.C. Principe, and T.F. Wong, Nonlinear Channel Equalization Using Multilayer Perceptrons with Information-Theoretic Criterion, *Proceedings of the International Workshop on Neural Networks for Signal Processing*, pp. 443-451, September 2001.

S.K. Patra, and Bernard Mulgrew, Fuzzy Techniques for Adaptive Nonlinear Equalization, *Signal Processing*, No.80, pp.985–1000, 2000.

Cheng Jian Lin and Wen Hao Ho, Blind Equalization Using Pseudo-Gaussian-Based Compensatory Neuro-Fuzzy Filters, *International Journal of Applied Science and Engineering*, Vol.2, No.1, pp.72–89, January 2004.

Jyh-Shing Roger Jang, ANFIS: Adaptive-Network-Based Fuzzy Inference System, *IEEE Transactions on Systems, Man, and Cybernetics*, Vol.23, No.3, pp.665–685, May/June 1993.

Nan Xie and Henry Leung, Blind Equalization Using a Predictive Radial Basis Function Neural Network, *IEEE Transactions on Neural Networks*, Vol.16, No.3, pp.709–720, May 2005.

J.S.R. Jang, and C.T. Sun, Functional Equivalence Between Radial Basis Function Networks and Fuzzy Inference Systems, *IEEE Transactions on Neural Networks*, Vol.4, No.1, pp.156–159, January 1993.

8

OFDM and Spatial Diversity

8.1 Introduction

In this chapter, we discuss the Orthogonal Frequency Division Multiplexing (OFDM) and Spatial Diversity techniques. Note that OFDM is one of the most recent and advanced techniques used in wireless mobile communications. OFDM is a Frequency Division Multiplexing (FDM) modulation technique for transmitting large amounts of digital data over a radio wave. OFDM works by splitting the radio signal into multiple smaller subsignals that are then transmitted simultaneously at different frequencies to the receiver. OFDM reduces the amount of crosstalk in signal transmissions. 802.11a WLAN, 802.16, and WiMAX technologies use OFDM.

The 4G cellular technology standard Long-Term Evolution (LTE) uses OFDM (Rappaport 1996). The high-speed short-range technology known as Ultra-Wideband (UWB) uses an OFDM standard set by the WiMedia Alliance. OFDM is also used in wired communications like power-line networking technology. One of the first successful and most widespread uses of OFDM was in data modems connected to telephone lines. ADSL and VDSL used for Internet access use a form of OFDM known as discrete multi-tone (DMT). And, there are other less well known examples in the military and satellite worlds.

The noise performance of OFDM was found to depend solely on the modulation technique used for modulating each carrier of the signal. The performance of the OFDM signal was found to be the same as for a single carrier system, using the same modulation technique. The minimum signal to noise ratio (SNR) required for BPSK was 7 dB, where as it was 12 dB for QPSK and 25 dB for 16PSK.

CDMA was found to perform poorly in a single cellular system, with each cell only allowing 7–16 simultaneous users in a cell, compared with 128 for OFDM. This was for a 1.25 MHz bandwidth and 19.5 kbps user data rate. This low cell capacity of CDMA was attributed to the use of nonorthogonal codes used in the reverse transmission link, leading to a high level of interuser interference.

The only main weak point that was found with using OFDM was that it is very sensitive to frequency and phase errors between the transmitter and receiver. The main sources of these errors are frequency stability problems, phase noise of the transmitter, and any frequency offset errors between the transmitter and receiver. This problem can be mostly overcome by synchronizing the clocks between the transmitter and receiver, by designing the system appropriately, or by reducing the number of carriers used (Leff 1994).

Coded Orthogonal Frequency Division Multiplexing (COFDM) is currently being used in several new radio broadcast systems including the proposal for high definition digital television, Digital Video Broadcasting (DVB), and Digital Audio Broadcasting (DAB). With CDMA systems, all users transmit in the same frequency band using specialized codes as a basis of channelization. The transmitted information is spread in bandwidth by multiplying it by a wide bandwidth pseudo random sequence. Both the base station and the mobile station know these random codes that are used to modulate the data sent, allowing it to de-scramble the received signal.

OFDM/COFDM allows many users to transmit in an allocated band, by subdividing the available bandwidth into many narrow bandwidth carriers. Each user is allocated several carriers in which to transmit their data. The transmission is generated in such a way that the carriers used are orthogonal to one another, thus allowing them to be packed together much closer than standard FDM. This leads to OFDM/COFDM providing a high spectral efficiency.

8.2 CDMA

CDMA is a spread spectrum technique that uses neither frequency channels nor time slots. With CDMA, the narrow band message (typically digitized voice data) is multiplied by a large bandwidth signal that is a pseudo random noise code (PN code). All users in a CDMA system use the same frequency band and transmit simultaneously. The transmitted signal is recovered by correlating the received signal with the PN code used by the transmitter.

Some of the properties that have made CDMA useful are

- Signal hiding and noninterference with existing systems.

- Anti-jam and interference rejection.

- Information security.

- Accurate ranging.

- Multiple user access.

- Multipath tolerance.

For many years, spread spectrum technology was considered solely for military applications. However, with rapid developments in LSI and VLSI designs, commercial systems also started to emerge.

8.2.1 Processing Gain of CDMA Systems

One of the most important concepts required in order to understand spread spectrum techniques is the idea of process gain. The process gain of a system indicates the gain or signal to noise improvement exhibited by a spread spectrum system by the nature of the spreading and despreading process. The process gain of a system is equal to the ratio of the spread spectrum bandwidth used to the original information bandwidth. Thus, the process gain can be written as

$$G_p = \frac{BW_{RF}}{BW_{sig}} \tag{8.1}$$

where BW_{RF} is the transmitted bandwidth after the data is spread, and BW_{sig} is the bandwidth of the information or data being sent.

8.2.2 Generation of CDMA

CDMA is achieved by modulating the data signal by a pseudo random noise sequence (PN code), which has a chip rate higher then the bit rate of the data. The PN code sequence is a sequence of ones and zeros (called chips), which alternate in a random fashion. Modulating the data with this PN sequence generates the CDMA signal. The CDMA signal is generated by modulating the data by the PN sequence. The modulation is performed by multiplying the data (XOR operator for binary signals) with the PN sequence. The basic CDMA transmitter is illustrated in Figure 8.1.

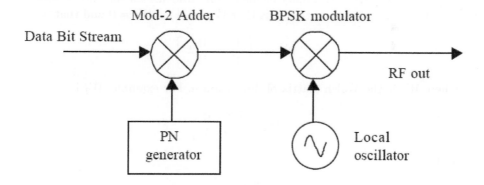

FIGURE 8.1
Simple Direct Sequence CDMA Transmitter.

The PN code used to spread the data can be of two main types. A short PN code (typically 10–128 chips in length) can be used to modulate each data bit. The short PN code is then repeated for every data bit, allowing for quick and simple synchronization of the receiver. Alternatively a long PN code can be used. Long codes are generally thousands to millions of chips in length, and thus are only repeated infrequently. Because of this they are useful for added security as they are more difficult to decode.

A typical direct sequence spread spectrum CDMA output signal for a binary sequence, $1, 0, 1, \dots$ is illustrated in Figure 8.2.

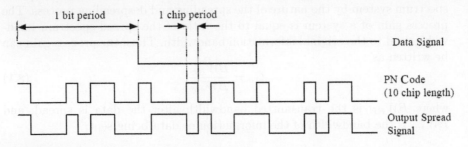

FIGURE 8.2
Direct Sequence CDMA Output Signal.

8.2.3   CDMA Forward Link Encoding

The forward link, from the base station to the mobile, of a CDMA system can use special orthogonal PN codes, called Walsh codes, for separating the multiple users on the same channel. These are based on a Walsh matrix, which is a square matrix with binary elements and dimensions that are a power of two. It is generated from the basis that $Walsh(1) = W_1 = 0$ and that

$$W_{2n} = \begin{bmatrix} W_n & W_n \\ W_n & \overline{W_n} \end{bmatrix}$$

where W_n is the Walsh matrix of dimension n. For example, W_2 is

$$W_2 = \begin{bmatrix} 0 & 0 \\ 0 & 1 \end{bmatrix}$$

and

$$W_4 = \begin{bmatrix} 0 & 0 & 0 & 0 \\ 0 & 1 & 0 & 1 \\ 0 & 0 & 1 & 1 \\ 0 & 1 & 1 & 0 \end{bmatrix}$$

Walsh codes are orthogonal, which means that the dot product of any two rows is zero. This is due to the fact that for any two rows exactly half the number of bits match and half do not. Each row of a Walsh matrix can be used as the PN code of a user in a CDMA system. By doing this the signals from each user are orthogonal to every other user, resulting in no interference between the signals. However, in order for Walsh codes to work the transmitted chips from all users must be synchronized. If the Walsh code used by one user is shifted in time by more than about 1/10 of a chip period with respect to all the other Walsh codes, it loses its orthogonal nature, resulting in interuser interference. This is not a problem for the forward link as signals for all the users originate from the base station, ensuring that all the signals remain synchronized.

8.2.4 CDMA Reverse Link Decoding

The reverse link is different from the forward link because the signals from each user do not originate from a same source as in the forward link. The transmission from each user will arrive at a different time, due to propagation delay and synchronization errors. Due to the unavoidable timing errors between the users, there is little point in using Walsh codes, as they will no longer be orthogonal. For this reason, simple pseudo random sequences are typically used. These sequences are chosen to have a low cross correlation to minimize interference between users. The capacity is different for the forward and the reverse links because of the differences in modulation. The reverse link is not orthogonal, resulting in significant interuser interference. For this reason the reverse channel sets the capacity of the system.

8.3 COFDM

COFDM is the same as OFDM except that forward error correction is applied to the signal before transmission. This is to overcome errors in the transmission due to lost carriers from frequency selective fading, channel noise, and other propagation effects.

In FDMA each user is typically allocated a single channel, which is used to transmit all the user information. The bandwidth of each channel is typically 10–30 kHz for voice communications. However, the minimum required bandwidth for speech is only 3 kHz. The allocated bandwidth is made wider than the minimum amount required to prevent channels from interfering with one another. This extra bandwidth is to allow for signals from neighboring channels to be filtered out, and to allow for any drift in the center frequency of the transmitter or receiver. In a typical system up to 50% of the total spectrum is wasted due to the extra spacing between channels. This problem becomes worse as the channel bandwidth becomes narrower and the frequency band increases.

Most digital phone systems use vocoders to compress the digitized speech. This allows for an increased system capacity due to a reduction in the bandwidth required for each user. Current vocoders require a data rate somewhere between 4 and 13 kbps, depending on the quality of the sound and the type used. Thus each user only requires a minimum bandwidth of somewhere between 2 and 7 kHz, using QPSK modulation. However, simple FDMA does not handle such narrow bandwidths very efficiently.

TDMA partly overcomes this problem by using wider bandwidth channels, which are used by several users. Multiple users access the same channel by transmitting their data in time slots. Thus, many low data rate users can be combined together to transmit in a single channel that has a bandwidth sufficient so that the spectrum can be used efficiently.

There are, however, two main problems with TDMA. There is an overhead associated with the change over between users due to time slotting on the channel. A change over time must be allocated to allow for any tolerance in the start time of each user, due to propagation delay variations and synchronization errors. This limits the number of users that can be sent efficiently in each channel. In addition, the symbol rate of each channel is high (as the channel handles the information from multiple users), resulting in problems with multipath delay spread.

OFDM overcomes most of the problems with both FDMA and TDMA. OFDM splits the available bandwidth into many narrow band channels (typically 100–8000). The carriers for each channel are made orthogonal to one another, allowing them to be spaced very close together, with no overhead as in the FDMA example. Because of this there is no great need for users to be time multiplex as in TDMA; thus there is no overhead associated with switching between users.

The orthogonality of the carriers means that each carrier has an integer number of cycles over a symbol period. Due to this, the spectrum of each carrier has a null at the center frequency of each of the other carriers in the system. This results in no interference between the carriers, allowing them to be spaced as close as theoretically possible. This overcomes the problem of overhead carrier spacing required in FDMA.

Each carrier in an OFDM signal has a very narrow bandwidth (i.e., 1 kHz); thus the resulting symbol rate is low. This results in the signal having a high tolerance to multipath delay spread, as the delay spread must be very long to cause significant intersymbol interference (e.g. $> 100~\mu$s).

8.3.1   OFDM Transmission and Reception

To generate OFDM successfully the relationship between all the carriers must be carefully controlled to maintain the orthogonality of the carriers. For this reason, OFDM is generated by first choosing the spectrum required, based on the input data and modulation scheme used. Each carrier to be produced is assigned some data to transmit. The required amplitude and phase of the carrier is then calculated based on the modulation scheme (typically differential BPSK, QPSK, or QAM). The required spectrum is then converted back to its time domain signal using an inverse Fourier transform. In most applications, an Inverse Fast Fourier Transform (IFFT) is used. The IFFT performs the transformation very efficiently, and provides a simple way of ensuring the carrier signals produced are orthogonal.

The Fast Fourier Transform (FFT) transforms a cyclic time domain signal into its equivalent frequency spectrum. This is done by finding the equivalent waveform, generated by a sum of orthogonal sinusoidal components. The amplitude and phase of the sinusoidal components represent the frequency spectrum of the time domain signal. The IFFT performs the reverse process, transforming a spectrum (amplitude and phase of each component) into a time domain signal. An IFFT converts a number of complex data points, of length that is a power of 2, into the time domain signal of the same number of points. Each data point in the frequency spectrum used for an FFT or IFFT is called a bin. The orthogonal carriers required for the OFDM signal can be easily generated by setting the amplitude and phase of each frequency bin, then performing the IFFT. Since each bin of an IFFT corresponds to the amplitude and phase of a set of orthogonal sinusoids, the reverse process guarantees that the carriers generated are orthogonal. The basic OFDM transmitter and receiver are illustrated in Figure 8.3.

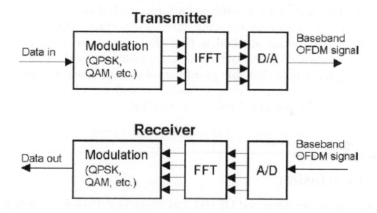

FIGURE 8.3
Basic OFDM Transmitter and Receiver.

8.3.1.1 Adding a Guard Period to OFDM

One of the most important properties of OFDM transmissions is their high level of robustness against multipath delay spread. This is a result of the long symbol period used, which minimizes the intersymbol interference. The level of multipath robustness can be further increased by the addition of a guard period between transmitted symbols. The guard period allows time for multipath signals from the pervious symbol to die away before the information from the current symbol is gathered. The most effective guard period to use is a cyclic extension of the symbol. If a mirror in time of the end of the symbol waveform is put at the start of the symbol as the guard period, this effectively extends the length of the symbol while maintaining the orthogonality of the waveform. Using this cyclic extended symbol, the samples required for performing the FFT (to decode the symbol) can be taken anywhere over the length of the symbol. This provides multipath immunity as well as symbol time synchronization tolerance (Bell et al. 1996).

As long as the multipath delay echoes stay within the guard period duration, there is strictly no limitation regarding the signal level of the echoes: they may even exceed the signal level of the shorter path! The signal energy from all paths just adds at the input to the receiver, and since the FFT is energy conservative, the whole available power feeds the decoder. If the delay spread is longer than the guard interval, then they begin to cause intersymbol interference. However, provided the echoes are sufficiently small they do not cause significant problems. This is true most of the time as multipath echoes delayed longer than the guard period will have been reflected off very distant objects.

Other variations of guard periods are possible. One possible variation is to have half the guard period a cyclic extension of the symbol, as above, and the other half a zero amplitude signal. Using this method the symbols can be easily identified. This possibly allows for symbol timing to be recovered from the signal, simply by applying envelop detection. The disadvantage of using this guard period method is that the zero period does not give any multipath tolerance; thus the effective active guard period is halved in length. It is interesting to note that this guard period method has not been mentioned in any of the research papers read, and it is still not clear whether symbol timing needs to be recovered using this method.

8.4   Conclusion

In this chapter, we discussed OFDM and frequency diversity techniques. The aim was to develop a mathematical model of the performance (BER) of OFDM verses channel noise. This was so that the simulated results could be verified, and to get a more in depth understanding of the transmission mechanism. The model developed is based on the transmission modulation technique being phase shift keying, and that the channel noise is Gaussian noise (i.e., white noise).

There are several processing stages required to generate and receive an OFDM signal. However, most of the processing is required in performing the FFT. The complexity of performing an FFT is dependent on the size of the FFT. The larger the FFT the greater the number of calculations required; however, since as the symbol period is longer the increased processing required is less than the straight increase in processing to perform a single FFT. It can be seen that because the symbol period increases with a larger FFT that the extra processing required is minimal.

The current status of the research is that OFDM appears to be a suitable technique as a modulation technique for high performance wireless telecommunications. An OFDM link has been confirmed to work by using computer simulations and some practical tests performed on a low bandwidth baseband signal. So far only four main performance criteria have been tested, which are OFDMs tolerance to multipath delay spread, channel noise, peak power clipping, and start time error. Several other important factors affecting the performance of OFDM have only been partly measured. These include the effect of frequency stability errors on OFDM and impulse noise effects. OFDM was found to perform very well compared with CDMA, with it outperforming CDMA in many areas for a single and multicell environment. OFDM was found to allow up to 2–10 times more users than CDMA in a single cell environment and from 0.7–4 times more users in a multicellular environment. The difference in user capacity between OFDM and CDMA was dependent on whether cell sectorization and voice activity detection were used (Magill 1994).

It was found that CDMA only performs well in a multicellular environment where a single frequency is used in all cells. This increases the comparative performance against other systems that require a cellular pattern of frequencies to reduce intercellular interference. One important major area which has not been investigated is the problems that may be encountered when OFDM is used in a multiuser environment. One possible problem is that the receiver may require a very large dynamic range in order to handle the large signal strength variation between users.

We have concentrated on OFDM; however, most practical systems would use forward error correction to improve system performance. Thus more work needs to be done on studying forward error correction schemes that would be suitable for telephony applications and data transmission (Gibson 1996).

Further Reading

T.S. Rappaport, *Wireless Communications, Principle & Practice*, IEEE Press, Prentice Hall, New York 1996.

B. Leff, Making Sense of Wireless Standard and System Designs, *Microwaves & RF*, pp. 113–118, February 1994.

T. Bell, J. Adam, and S. Lowe, Communications, *IEEE Spectrum*, pp. 30-41, January 1996.

D. Magill, Spread-Spectrum Technology for Commercial Applications, *Proceedings of the IEEE*, Vol. 82, No. 4, April 1994.

J. D. Gibson, *The Mobile Communications Handbook*, CRC Press, Boca Raton, FL, pp. 366-368, 1996.

9

Conclusion

9.1 Introduction

The research carried out for this work primarily examines the design and analysis of Neuro-Fuzzy Adaptive filters for ISI mitigation in mobile cellular channels. We start our discussion with a brief introduction to the principles of mobile cellular telephony. We then review the equalizer models currently available. The proposed equalizers for Nonlinear Time-Variant (NLTV) channels are discussed. The modeling of UWB channels based on the Channel Covariance Matrix (CCM) is undertaken. It is also shown that the ANFIS equalizer can be suitably adapted for UWB as well. Then, we arrive at a generic framework for the ANFIS, CNFF and FAF based adaptive equalizers. We show that all the three can be brought under variants of RBF neural networks. We propose a new modular approach for equalizers for NLTV channels.

The signal transmitted through a channel suffers from linear, nonlinear, and additive distortion. The conventional method for compensation of channel distortion is based on introducing a linear equalizer (linear inverse filter to the channel frequency response) to the output of the channel. This design methodology is appropriate when the channel model is precisely known and the characteristics of the channel are time invariant. When the channel has time-varying characteristics, adaptive equalizers are used. Various approaches have been used for nonlinear channel equalization. Classical approaches are based on the knowledge of the parametric channel model. The next type is the decision feedback equalizer that improves the performance of the equalizer. Nowadays neural networks are widely used for channel equalization. One of the classes of nonlinear adaptive equalizers is based on Multi-Layer Perceptrons (MLP) and RBF. MLP equalizers require long training and are sensitive to the initial choice of network parameters. RBF equalizers are simple and require less time for training, but they usually require a large number of centers, which increases the complexity of computation.

Mobile cellular channels are generally considered as nonlinear and time variant. They also show *Rayleigh fading* or *Ricean fading* characteristics. The fading characteristics will be those of a *Ricean distribution* if, apart from the major ray, one more component reaches the receiver (*two-ray model*). It will exhibit a *Rayleigh distribution* if three or more multipath components reach the receiver (*three-ray model*).

9.2 Major Achievements of the Work

The major achievements of this work can be summarized as follows:

- The mobile cellular channel can, in general, be modeled as an NLTV with *Rayleigh* or *Ricean* fading characteristics. However, under limiting conditions, it can also be modeled as a linear time-variant channel. It is shown that the indoor mobile cellular channel has *Rayleigh* fading characteristics, using a *three-ray model*. It is also shown that the mobile cellular channel is either Linear Time-Variant (LTV) or NLTV.

- The mobile channel being nonlinear, nonlinear equalizers are more appropriate for them. We consider three such equalizers—FAF, ANFIS, and CNFF. The performances of these are studied. Various structures of the ANFIS-based equalizers are considered for channel equalization and their performances are compared. It is also shown that equalizers based on the ANFIS structure can be adapted for UWB channels as well. Consequent to the deployment of UWB in most modern communication scenarios, the need for equalization at these frequencies has gained more momentum.

- An RBF neural network framework for the above three equalizer models is derived. This is especially useful for easier comparative performance analysis of the equalizers. It is shown that the lower order ANFIS-based equalizer (ANFIS–25 with 75 nodes and 25 rules) has almost identical performance of that of an RBF neural network based equalizer.

- A modular approach is proposed for the design and simulation of equalizers for nonlinear time-variant channels. In this model, a nonlinear prefilter precedes the equalizer block at the receiver. This approach provides considerable improvements in equalizer performance. It is shown that the method is highly efficient in removing higher order nonlinearities introduced by the nonlinear channel. The prefilter is implemented using an ANFIS–37 structure and RBF neural network. Both of them perform equally well in removing the nonlinearities at the channel output. It simplifies the design and simulation of channel equalizers where there is nonlinearity due to output power amplifier performance and due to the channel itself.

9.3 Limitations of the Work

The principles discussed in this monograph are equally applicable to all kinds of mobile cellular systems including GSM and CDMA based technologies. We had considered only a few of the currently available modulation schemes in digital communication in this work. It is imperative that the principles evolved in this work be extended to several other modulation schemes like 256QAM or 512QAM.

The novel modular approach in the design of equalizers introduced in Chapter 6 needs to be analyzed more critically. There are some recent papers in that direction by some authors.

Even though we have considered many techniques for channel equalization of mobile cellular channels, practical implementation of the algorithms is not considered. As seen from the literature, most of the algorithms are suitable for implementation on DSP platforms (Ahmed et al. 2004).

9.4 Scope for Further Research

To conclude the monograph, the following are some pointers for further research work which can lead to interesting results:

- Possible extensions of this work can be found useful in developing equalizers for MIMO systems. Multi-channel CDMA is one such application. This is a fast developing area of research.

- As we can see from current literature, wireless access is another highly investigated topic of intense research activity. Most of the principles developed can be used in wireless networking as well.

- The equalizers based on the ANFIS structure can be extended for equalization of UWB channels, as shown in Chapter 4. Further, it can be exploited in the deployment of Personal Area Networks (PAN) and Body Area Networks (BAN).

- Broadband wireless technology will have an important role in the future evolution of advanced global telecommunications (Ariyavisitakul and Li 1998, Cosovic et al. 2005). The IEEE 802.11 standard comprehensively covers data transmission in wireless LANs, which includes methods for CCI suppression and equalization (Luo and Liu 2002).

- As wireless LANs are getting popular rapidly, many predict that wireless LANs can be used to build a wireless Internet and compete against

3G systems in terms of providing broadband wireless data service at *hot spots*. This scenario can be exploited to the full extent. More study in this direction is most appropriate.

Further Reading

Yaser Ahmed, Farooq Jawed, Sohaib Zia, and Muhammed Sheraz Aga, Real-Time Implementation of Adaptive Channel Equalization Algorithms on TMS320C6x DSP Processors, *Proceedings of E-Tech 2004*, pp.101–108, July 2004.

Sirikiat Ariyavisitakul and Ye Geoffrey Li, Joint Coding and Decesion Feedback Equalization for Broadband Wireless Channels, *Proceedings of the 1998 Vehicular Technology Conference, VTC98*, Vol.3, pp.2256–2261, May 1998.

Ivan Cosovic, Michael Schnell, and Andreas Springer, Combined Equalization for Uplink MC-CDMA in Rayleigh Fading Channels, *IEEE Transactions on Communications*, Vol.53, No.10, pp.1609–1614, October 2005.

Hui Luo and Ruey-Wen Liu, Apply Autocorrelation Matching Method to Outdoor Wireless LAN on Co-Channel Interference Suppression and Equalization, *Proceedings of the 2002 IEEE Conference on Wireless Communications and Networking–WCNC2002*, pp.459–464, Vol.1, March 2002.

Index

Printed and bound by CPI Group (UK) Ltd, Croydon, CR0 4YY

18/10/2024

01776262-0003